Hunting Plants

Hunting Plants

The story of those who discovered
the flowering plants and ferns
of North Lancashire

Eric Greenwood

SCOTFORTH

First published in 2015 on behalf of the author by
Scotforth Books (www.scotforthbooks.com)
ISBN   978-1-909817-22-7
Typesetting and design by Carnegie Book Production, Lancaster.
Printed in the UK by Jellyfish Solutions

Contents

Acknowledgements

I am very grateful to all the libraries, record offices, herbaria and museums that have allowed access to their collections. Many individuals and relatives of the deceased botanists have help source photographs and provided personal information whilst others have given helpful advice in developing my ideas for the book. These include:

Melissa Atkinson and Tabitha Driver (Religious Society of Friends); Wendy Atkinson (World Museum, Liverpool); Tom Blockeel, Rachel Carter and Phil Stanley (British Bryological Society); Sam Bosanquet; Margaret Bromley-Webb; John Calman; Anne Catterall (Bodleian Library, Oxford); Jerry Cooper; John Edmondson; Richard and Mavis Gulliver; Margaret Haggis; George Heaton; John Hodgson; Andrew Holgate (University of Lancaster Library); Jim Jarvis; Peter Jepson; David Livermore; Clive Lovatt; Heather McAvoy-Marshall; Serena Marner (Department of Plant Sciences, Oxford University); Geoff Morries; David and Siobhan Newton; Katherine Slade and Sally Whyman (National Museum of Wales); Jeremy Steeden and William Wheldon.

Special thanks go to my wife, Barbara for her forbearance, encouragement, IT skills and detailed proofing of the script.

In bringing the book to fruition I am grateful to Anna Goddard, Lucy Frontani and Penny Hayashi of Carnegie Publishing.

Eric Greenwood
September 2014

How the plants were found

Introduction

One of the fascinating aspects of compiling a county or local flora is researching who found the plants and what inspired them to make records. Very often flora writers included a section on the contributors (e.g. Halliday, 1977). Occasionally separate publications have been written (e.g. Coles, 2011; Hodkinson and Steward, 2012) whilst David Allen (1976, 1986, 2001) published books covering the whole country.

After nearly 50 years work it was felt that the priority was to publish the *Flora of North Lancashire* (Greenwood, 2012). To include a section on the contributors would not only increase the size of an already large volume but it would also require further research. Accordingly the story of who recorded the North Lancashire flora is presented here.

Several writers have been interested in the role of artisan naturalists in exploring the flora of Lancashire (e.g. Percy, 1991; Secord, 1994a, 1994b). However, although much has been written about their activities in the late eighteenth and early nineteenth century, the artisan naturalists are mostly involved with the developing cotton industry in the Mersey Basin and particularly in the Manchester area. North of the R. Ribble artisan naturalists do not seem to have been involved to the same extent although they probably contributed to the unpublished and lost manuscript (if it was ever written) of 'Flora Prestoniensis'. Nevertheless the story is no less interesting especially in the role and influence of the Religious Society of Friends in Britain (Quakers) that continues today.

The early years to the middle of the nineteenth century

The earliest records of North Lancashire plants are attributed to John Gerard (1636) who recorded Cloudberry (*Rubus chamaemorus*) from Gragareth, where it is still found. However many of the earliest records from the seventeenth and eighteenth centuries were based on those made by John Ray (1627–1705) or more usually by his correspondents and cited by him in his publications. It is believed that Ray made one visit to Lancashire (Plate 1) in 1660 as part of his northern tour (Raven, 1942). After visiting the Isle of Man he returned to England at the mouth of the R. Wyre in Lancashire. At that time there were two ports: Skippool on the south bank and Wardleys (Plate 2) on the north bank (Crosby, 1998). As the few records he made on that trip are from Pilling it is suggested that he disembarked at Wardleys and took a route along a ridge of slightly higher ground. This separated the large undrained bogs and fens of Pilling and Rawcliffe Mosses before reaching the main north–south road through Lancashire at Garstang. On this journey he recorded Marsh Fleawort (*Tephroseris palustris*), familiar to him in Cambridgeshire and Hare's-tail Cottongrass (*Eriophorum vaginatum*) both from Pilling (Ray, 1670).

The Quaker connection

George Fox (1624–1691) founded the Quakers in the Furness area of Lancashire, now part of Cumbria. Amongst his converts was Thomas Lawson (1630–1691). In 1650 Lawson was at Cambridge University training to be a priest and was thus a near contemporary of John Ray. He went as clergyman to Rampside in Furness and in 1652 heard George Fox preach. He subsequently gave up his living and became a Quaker (Whittaker, 1986).

For the next 100 years or so such botanical exploration as took place in North Lancashire owes much to the Quakers. Indeed this involvement continues to the present day and at times they have taken a lead role, e.g. Albert Wilson (see below). The philosophy of the Quakers encouraged enquiry (seeking the truth), education and meticulous record keeping. Many kept diaries and some of these have survived. Their doctrine encouraged travel, not only in Britain but further afield as well. Not being part of the established church they were persecuted and suffered hardship. For the most part they formed a small, close-knit society of friends and relatives offering each other mutual support with a particular stronghold in north western England but extending to all parts of the country and overseas, especially America (Raistrick, 1968).

Thomas Lawson became a prominent Quaker, travelling and preaching widely. He settled at Great Strickland in Cumbria where, when he was able, he ran a school. In 1674 he was known at Swarthmore (in Furness), the home of George Fox and his wife Margaret Fell (the widow of Judge Fell), as a man qualified to instruct in herbs. At about the same time his own notebooks show that he became interested in field botany and all his records are post 1674 (Whittaker, 1986). Thomas Lawson is best known for his contribution to Cumbrian botany but at least eighteen of his records are attributed to West Lancaster, V.C. 60 where, for a short period, he taught at Warton (Marshall, 1967). It is not known when Ray and Lawson first became acquainted but following a letter from Lawson to Ray (Lankester, 1848) written in 1688, Lawson's records appear in Ray's publications.

After this initial burst of enthusiasm for exploring the natural world the middle part of the eighteenth century at least was a barren period throughout the country (Allen, 2001). The situation in north western England (Halliday, 1997) and North Lancashire was little different.

Nevertheless after a short break the names of Quakers and their friends re-appear in the botanical literature. John Wilson (1696–1751) of Longsleddale near Kendal published a record for Mossy Saxifrage (*Saxifraga hypnoides*) from Ease Gill Kirk, Leck (Wilson, 1744) where it is still found abundantly (Plate 3). It is not clear if John Wilson was a Quaker or not but it is possible as in later life he moved to Newcastle-upon-Tyne. Here he was associated with the Quaker Isaac Thompson, surveyor and printer. Unfortunately Quaker records reveal other John Wilsons from Kendal at this time and it is difficult to distinguish the individuals with the same name and similar dates of birth and death.[1]

Towards the end of the century another Quaker contributed significantly to the documentation of the North Lancashire flora. In 1775 James Jenkinson published his *A generic and specific descriptions of British Plants...* in which he recorded the localities of 62 species of flowering plants and ferns in North Lancashire. Most of his records were from the Silverdale and Warton areas but a few were from further afield, especially Middlesex. From a single specimen (Plate 4) at Friends House, London it appears he formed an herbarium or *Hortus Siccus* but this is presumed lost.[2]

4

Family Relationships of Quaker Botanists

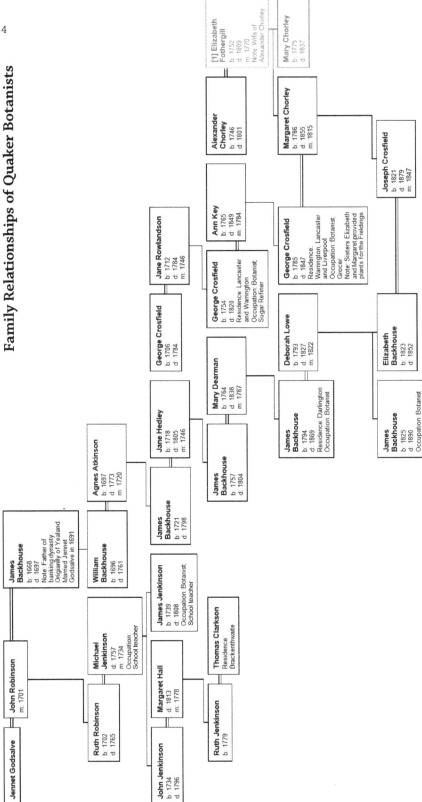

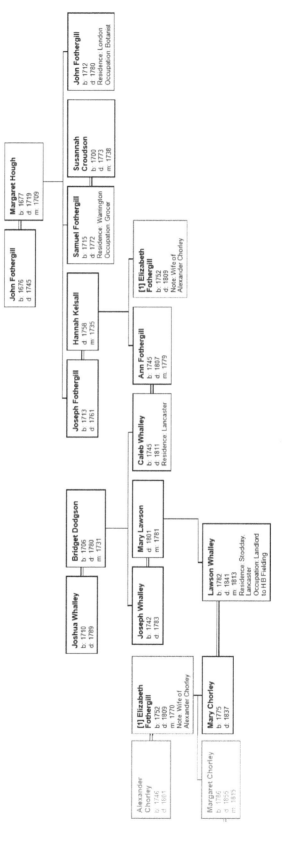

John Fothergill
b 1676
d 1745

Margaret Hough
b 1677
d 1719
m 1709

John Fothergill
b 1712
d 1780
Residence: London
Occupation: Botanist

Susannah Croudson
b 1700
d 1773
m 1738

Samuel Fothergill
b 1715
d 1772
Residence: Warrington
Occupation: Grocer

Hannah Kelsall
d 1758
m 1735

[1] Elizabeth Fothergill
b 1752
d 1809
Note: Wife of Alexander Chorley

Joseph Fothergill
b 1713
d 1761

Ann Fothergill
b 1745
d 1807
m 1779

Caleb Whalley
b 1745
d 1811
Residence: Lancaster

Bridget Dodgson
b 1706
d 1780
m 1731

Joshua Whalley
b 1710
d 1789

Mary Lawson
d 1801
m 1781

Joseph Whalley
b 1742
d 1783

Lawson Whalley
b 1782
d 1841
m 1813
Residence: Stodday,
Lancaster
Occupation: Landlord
to H B Fielding

[1] Elizabeth Fothergill
b 1752
d 1809
m 1770
Note: Wife of Alexander Chorley

Mary Chorley
b 1775
d 1837

Alexander Chorley
b 1746
d 1801

Margaret Chorley
b 1786
d 1855
m 1815

By the end of the eighteenth century the 'Age of Enlightenment' was fully established and documentation is available showing friendships and relationships. Quakers through their weekly, monthly and yearly meetings have a system enabling individual members to meet others in their area and elsewhere in the country. As a consequence there was an interchange of thoughts and ideas and as these developed people influenced one another. Thus, whilst James Jenkinson was at Yealand Conyers one of his father's pupils[3] was James Backhouse (1721–1798). He was the grandson of James Backhouse (1668–1697) who was the second husband of Janet Godsalve. She had previously been married to John Robinson and was the grandmother of James Jenkinson (see Family Relationships of Quaker Botanists chart, pp. 4–5). James Backhouse (1721–1798) went to live in Darlington and founded a dynasty of Backhouses famous for their commercial enterprises and botanical pursuits (Davis, 1996).

Until recently little was known about James Jenkinson. Desmond (1994) gives the briefest of entries whilst a note by L. Petty (1902) provides details of a headstone marking his grave in Yealand Conyers. The headstone survives in a well-hidden copse (Plate 5).

Records at Friends House[1] confirm that James Jenkinson was a Quaker born at Yealand Conyers on 6 November 1738 and died on 15 October 1808. His parents were Michael Jenkinson and Ruth Robinson. Michael Jenkinson was the local Quaker school master appointed in March 1729[4] to which position his son James succeeded. James had seven brothers and sisters. The eldest brother was John, described as Yeoman, and from whose property in Yealand James recorded plants. However both James and John were prominent members of the Yealand and Lancaster Meetings. The enclosure award for Yealand Common in 1778[5] shows both John and James as landowners and along with others acted on behalf of the Yealand Meeting of Friends that were allotted part of the Common. John was also known as a school master[4] but he was better known for his proposals for reclaiming part of Morecambe Bay, agricultural improvements generally (Holt, 1795) and proposals for a canal (Hadfield and Biddle, 1970). John and James also seem to have been involved in the printing trade as a pamphlet at Lancaster Reference Library is printed by J. and J. Jenkinson of Yealand. A copy of James Jenkinson's book inscribed John Jenkinson is also at the library.

Archives of the Lancaster Monthly Meeting[6] show that James Jenkinson left the Yealand Meeting for London in 1764 returning from the Horsby Down Meeting in 1768. It is not known if he travelled else-where although some records, possibly his, are from Suffolk and it seems that apart from this absence he lived at Yealand all his life. He died unmarried at Brackenthwaite, Yealand Redmayne, the home of his niece, Ruth Clarkson, on 15 October 1808. His grave is described as being in a Croft behind his house at Yealand Conyers. This house is known today as Green Garth.[4]

One of the most famous of the eighteenth century Quaker botanists was Dr John Fothergill (1712–1780) whose family came from Yorkshire but he lived most of his life in London (Raistrick, 1968, Corner and Booth, 1971; Fothergill, 1998). It is surely inevitable that James Jenkinson met Fothergill during his time in London and it is from this period that James Jenkinson appears to have become interested in botany.

The Fothergill family had further connections with North Lancashire botany. John Fothergill's brother Samuel (1715–1772) and his wife had a shop in Warrington selling a variety of goods including tea. However, Samuel and his wife were also guardians of Dr J.C. Lettsom (1745–1815) when he was a boy and whilst his parents ran a sugar plantation on the British Virgin Islands in the Caribbean. John Lettsom was born on the island of Little Jost Van Dyke and came to England for his educa-tion on one of the Rawlinson family's ships trading between the Virgin Islands and Lancaster. The Rawlinsons were well-known Quaker traders in Lancaster (David and Winstanley, 2013). Later Lettsom served his apprenticeship in apothecary before being introduced to Samuel's brother John. John Lettsom subsequently went on to pursue a distinguished botanical career of his own but amongst his better known works was his *Natural History of the Tea Tree* published in 1772. Was his interest in tea stimulated by his guardian's business interest in tea?

Whilst James Jenkinson probably met Dr John Fothergill in London he must have met the Crosfield family in Kendal or Lancaster when they attended the Lancaster Meeting[6]. The Crosfields originated in the Beetham and Kirkby Lonsdale areas of north western England. George (1754–1820), son of George (1706–1784) was born in Preston Patrick near Kendal and aged sixteen was apprenticed to a grocer in Kendal. His father was a Yeoman and according to J.F. Crosfield (1980) helped his friend Samuel Fothergill in Warrington by allowing his son to manage Fothergill's Warrington shop in 1777[7] and which shortly afterwards he owned (Anon., 1792). There is some confusion over the details as Samuel

Fothergill died in 1772[8] and it is likely that George Crosfield (1754–1820) came to run the shop after his death. The business in Warrington prospered and after a few years George Crosfield (1754–1820) added a wholesale business and later moved to Lancaster. Here he ran a sugar refinery when at that time most of the sugar imported into Lancaster came from the British Virgin Islands. Inevitably this included sugar from the Lettsom estate and from other Lancaster Quakers including the Rawlinson, Lawson (who had also had a sugar refinery in Lancaster) and Chorley families (David and Winstanley, 2013). Amongst the Lancaster traders and merchants at that time was John Simpson who, no doubt, also brought sugar and other commodities to Lancaster from the British Virgin Islands (Dalziel, 1993).

When George Crosfield (1754–1820) was in Warrington it is likely that in the few years between 1777 and 1780 when Dr John Fothergill died he met the great botanist when he spent the summer at his Cheshire home, Lee Hall. Many years earlier his brother Samuel had provisioned the house. However there is no doubt that the Crosfield and Fothergill families were well known to each other through friendship and marriage (Crosfield, 1843; Crosfield, 1980). Perhaps it was this connection as well as with James Jenkinson that influenced the Crosfields to become amongst the early recorders of the North Lancashire flora. George Crosfield (1754–1820) also had a son called George (1785–1847) who took over the Warrington business from his father and also became a botanist and cultural figure in the town. With his father he also explored the flora of the Lancaster area.

Not much evidence for the Crosfields' botanical exploration of the Lancaster area exists but 24 surviving herbarium specimens were acquired by C.E. Salmon (1912) and later deposited at the Natural History Museum in London. They collected their plants between 1782 and 1814 and their significance lies in the vouchers for the wet heath formerly covering Lancaster Moor, now a park (Plate 6). Lancaster Moor formed the western edge of the Bowland Fells and the plants they found included Parsley Fern (*Cryptogramma crispa*) growing at a very low elevation. Their surviving collections also provide voucher specimens for plants long lost from the Mersey valley and the Warrington area.

The business interests of the Crosfield family extended into wire and soap making in Warrington and particularly through tea and the grocery business became Harrison (another Quaker family) and Crosfield in 1844 and in 1916 Twining Crosfield Ltd before being absorbed into British Associated Foods in 1964.

Through the younger George Crosfield (1785–1847) the influence of the Crosfield family in North Lancashire botany was to continue. He married Margaret Chorley in Lancaster in 1815, who shortly before her marriage came to live with her sister Mary, who had married Lawson Whalley MD (1782–1841) in 1813. He owned much of Stodday (Plate 7), a parish to the south of Lancaster.[9] He was a prominent Lancaster citizen and Justice of the Peace (Howson, 2002). However in tracing the relationships of the Chorley family it is found that John Fothergill was Margaret and Mary Chorley's grandfather. Furthermore George and Mary Crosfield's son, Joseph (1821–1879) married Elizabeth Backhouse, daughter of James Backhouse (1757–1804) of Darlington. Thus all these eighteenth and early nineteenth century Quaker botanists were either related or acquainted with one another. See Family Relationships of Quaker Botanists chart, pp. 4–5.

Also associated with Lancaster in the first half of the nineteenth century were Samuel Simpson (1802–1881), Henry Borron Fielding (1805–1851) and his wife Maria née Simpson (1805–1895). The literature shows that H.B. Fielding was a well-known if retiring botanist who assembled an important collection that formed the basis of the Fielding-Druce Herbarium at Oxford University. There are references to his artist wife Maria who illustrated her husband's publications and to their exploring the botany of the area where they lived in Lancashire.[10] Until recently only a few references to their Lancashire work and the odd specimen at Oxford seem to have survived. However also at Oxford and in the Watson Herbarium at the Royal Botanic Gardens, Kew rather more specimens were found collected by Samuel Simpson, brother of Maria Fielding. Some of Samuel's records are cited in the literature and he wrote a few botanical notes but his activities were confined to a few years around 1840. Neither Simpson nor Fielding was mentioned by Wheldon and Wilson (1907) and almost nothing was known about Mrs Fielding.

Following genealogical, archival and literature research it has been possible to elucidate the background to the Simpson and Fielding families and their links to Lancaster and the Quakers. However both the Simpsons and the Fieldings were of the established Church of England.

On the paternal side the Simpsons were merchants in Lancaster originating from Yeomen in the Furness area of Lancashire and also involved in the Lancaster–West Indies trade (Elder, 1992). In particular Samuel Simpson's father, John, was described as a merchant in the West Indies trade. On his mother's side there were connections to textile mills at Wray in the Lune valley. Samuel Simpson's grandfather had married Maria Salisbury and this connection involved prosperous merchants, lawyers and Bowland (Newton-in-Bowland) farmers. The Salisburys were also related to other well-known families in the area, e.g. the Wallings and Dawsons of Warton (Mourholme Local History Society, 1998). Thus a picture emerges of a prosperous family well-known in North Lancashire and Bowland with friendships and relationships with similar families in the region.

The Fieldings came from a different background of more humble beginnings in the Church and Pendle areas of South Lancashire. Henry Borron's paternal grandfather, Joseph Fielding (1728–1802), was a small-scale cotton manufacturer in Church but also had a warehouse in Manchester. The epitaph to his grave at Church reads:

> To the memory of Joseph Fielding of Hippings who departed this life
> Sept 16th 1802 aged 74 years.
> Just of his word in every thought sincere
> An honest Man, a faithful friend lies here.
> Good Sense and Reason did his actions guide
> He lived respected and lamented Dy'd

suggesting he was respected by the community.

To what extent his business interests competed or conflicted with those of the nearby Peel family are unclear. Nevertheless by 1778 Joseph's eldest son Henry (1757–1816), aged 22, was able, in partnership with Peter Sharples, to build a calico printing business at Catterall.[11] According to a 'Memoir' (Plate 8) at Oxford University[10] he was able to do this with a legacy from his maternal grandfather, a Dr Hartley MD. The identification of neither Peter Sharples nor Dr Hartley has been elucidated as the birth of Henry's mother, Ann (c. 1732–1803) has not been discovered. Also of significance is Henry Fielding's father-in-law, James Borron (1726–1804). He was a very wealthy businessman trading in cotton in Manchester with large warehouses. It was no doubt through these business interests that Joseph Fielding and James Borron got to know each other and that their children Henry Fielding and Susanna Borron met and in due course married. Susanna brought with her a substantial dowry, which was probably invested in the calico printing works at Catterall. By 1800 the business interests in calico printing etc. of the Peel and Fielding families were competing to be amongst the largest in the country. However their future fortunes were to be very different. The Peel family went on to ever greater fortunes with Sir Robert Peel (1788–1850), Prime Minister and creator of the modern Police force, perhaps their most famous member. On the other hand the Fieldings were to disappear into obscurity but not before they too had made a fortune at the works at Catterall.

Henry Borron Fielding (1805–1851) was born at Myerscough, near Catterall but lived at his parents' house in the fashionable Winckley Square in Preston. He always suffered from ill health and was privately educated. However disaster hit the family when in 1816 his father Henry died. Henry was then a Deputy Lieutenant for Lancashire and was a wealthy man owning several small estates in Lancashire besides his house at 1 Winckley Square (Plate 9). He was therefore able to provide for his wife and children and when his widow died enabled his son to lead the life of a gentleman without any necessity to earn a living. The business was left to Henry's brothers and whilst it continued to prosper for a short while the business climate changed and the company went spectacularly bankrupt in 1831 (Russell, 1986). This caused immense distress to the many employees.

Henry Borron Fielding took no interest in his uncles' business at Catterall and although he had no need to follow a career his mother was keen that he should be well-educated and qualified in a suitable profession. Thus he was sent to Liverpool to train as a solicitor.

By the 1820s Liverpool was one of the most important trading ports in the world. During the eighteenth century it had grown enormously and was a large cosmopolitan city. The importance of that trade and the resultant poverty in parts of the city are often areas of discussion (Belcham, 2006) but trade also brought wealth. In the early nineteenth century Liverpool attracted a large professional class and with them developed a vibrant cultural sector. A flavour of this cultural and professional life can be seen in the activities of the Rathbone family (Nottingham, 1992) whose banking business flourishes today. Perhaps it was not surprising that the frail Henry Borron Fielding came to Liverpool to train as a solicitor. Equally it was not surprising that Samuel Simpson also came to the city to train as a solicitor and that his sister, Maria, came to train as an artist. It was in Liverpool that the Fieldings met the Simpsons.

The 'Memoir'[10] suggests that H.B. Fielding's interest in botany and natural history owes its origin to his childhood at Myerscough, where the library and garden were an inspiration to him. However much of his time was spent in Preston where the Shepherd library, now in the Harris Museum, Preston, further encouraged him in his interests. Then, following the death of his mother in 1824 (not 1825 as in the 'Memoir') and marriage to Mary Maria Simpson in 1826, Fielding moved to Stodday Lodge (now part of Lunecliffe Hall) to the south of Lancaster (Plate 7) without qualifying as a solicitor. Here he rented the property from his neighbour Lawson Whalley.[9] A neighbouring estate was owned by Edward Dawson (1794–1876), a distant cousin of Mrs Fielding. Meanwhile Samuel Simpson qualified as a solicitor and set up in business on Castle Hill in Lancaster city centre (Plate 10).

The Fieldings, Lawson Whalley, Edward Dawson, George Crosfield (1754–1820) and Samuel Simpson were all members of the Lancaster Literary, Scientific and Natural History Society between 1836 and 1849.[12] Simpson was secretary throughout and Dawson served as President, 1836–1837. A Miss Crosfield was also a member for part of the time. This is probably Margaret or Elizabeth who married Richard Bull of Yealand in 1843, sisters of George. During this period Samuel Simpson was actively pursuing his botanical interests as well as practising as a solicitor in Lancaster. He was not only secretary of the Society but gave lectures and donated specimens and books to the Society's museum. On one occasion Samuel's brother, Richard Salisbury Simpson, also a botanist (Desmond, 1994), contributed to the Society's museum whilst H.B. Fielding contributed regularly but gave only one lecture on 27 January 1837. According to the 'Memoir' this was the only lecture he gave anywhere. Another contributor was John Just (1797–1852) who, until 1832, was teaching at Kirkby Lonsdale. He later moved to Bury where he became a prominent botanist and teacher. He produced a list of the plants in the Kirkby Lonsdale area but it appears it was subsequently lost.[13] At Stodday the Fieldings devoted their lives to the study of botany and natural history. Early on H. B. Fielding saw a collection of dried plants at the home of a friend and decided to form a collection of his own.[10] It is tempting to think that this friend was the Quaker George Crosfield who had an herbarium at this time.

Only fragments of the contributions made by Samuel Simpson and the Fieldings remain. The large herbarium that H.B. Fielding assembled contains almost nothing collected by him and apart from a few Simpson gatherings there is nothing from Lancashire. With the Simpson specimens in Watson's collection at the Royal Botanic Gardens, Kew and elsewhere it is clear that he must have gathered a substantial collection. However Simpsons's original collection has not been found and is presumed lost.

A further source of records is contained in six volumes of paintings by Mrs Fielding with a commentary by her husband.[15] Some of the plants were provided by the Misses Crosfield, possibly Elizabeth (see above), Margaret or Hannah who married James Ryley of Yealand Conyers in 1835, daughters of George Crosfield (1754–1820). These are perhaps the most interesting records as they provide additional detail on weeds and farming practices of the day (Plate 11).

These sources reveal that the Fieldings' botanical activity was limited to a few years between 1830 and 1839. This covers the period after Henry Borron Fielding and Mary Maria were married and were living at Stodday but ceases when they moved to Bolton Lodge at Bolton-le-Sands in 1839. Samuel Simpson's botanical activity covers much the same period when he was living at Castle Hill.[14] However in 1844 his father died and in the same year he married Ann Atkinson, the daughter of Richard Atkinson of Ellel Grange. His marriage linked him with notable Prestonians. Anne's mother was the daughter of Nicholas Grimshaw (1757–1838), Mayor of Preston on three occasions two of which were as Guild Mayor, a particularly prestigious appointment (Crosby, 2012). Also Anne's brother Richard Atkinson-Grimshaw was vicar of Cockerham. Through his wife Samuel Simpson probably had access to substantial funds. The marriage appears to have been important in two ways. In 1844 Samuel built a large house then on the outskirts of Lancaster, which he called The Greaves. Today it is a restaurant known as Greaves Park. Secondly within a few years he took holy orders and in 1851 he became Chaplain at St Thomas', Douglas, Isle of Man. His botanical interests ceased around the date of his marriage. Thus the botanical interests in Lancashire of both the Fieldings and Samuel Simpson (Plate 12) coincided with the period when the Lancaster Literary, Scientific and Natural History Society flourished.

Although the records are limited to a short period (1830s and early 1840s) they provide sufficient information to describe some of the interesting areas near Lancaster. In particular Lancaster Moor but of special interest was the specimens painted or gathered from a floristically rich field they called the 'Fairy Field' near Stodday. The flora included Bird's-eye-Primrose (*Primula farinosa*), which today is hard to believe. Other plants of the 'Fairy Field' were Pyramidal Orchid (*Anacamptis pyramidalis*), Marsh Helleborine (*Epipactis palustris*), Common Spotted-orchid (*Dactylorhiza fuchsii*), Grass-of-Parnassus (*Parnassia palustris*) and Common Butterwort (*Pinguicula vulgaris*) [Plate 11]. Today, if the field concerned has been correctly identified, it is improved grassland, probably MG6 *Lolium perenne-Cynosurus cristatus* grassland with few associated species (Rodwell, 1992). Another of Samuel Simpson's discoveries was that of Marsh Gentian (*Gentiana pneumonanthe*) near Morecambe. The present day urban sprawl gives little indication that less than 200 years ago there was an extensive wet heath in the area. On the other hand nothing is mentioned of the Lancaster Canal. In the 1830s was it truly devoid of aquatic and marginal vegetation? Within twenty years Ashfield (see below) demonstrated that this was not the case.

A further member of the Lancaster Literary, Scientific and Natural History Society was William Hall who, from 1839 to 1849, was one of the curators of the society's collections. He was probably William Hall FRCS (1817–1891), a prominent Lancaster citizen and medic becoming Mayor of Lancaster in 1878–1879 and holding various public offices in the city. However it was not until the 1990s that a small collection of flowering plants collected by a William Hall in 1862 was found at Lancaster University and transferred to World Museum, Liverpool. It seems likely that the collection belonged to William Hall (1817–1891) although it was assembled some years after the heyday of the Lancaster Society. Alternatively it could have been formed by some other unidentified person of the same name. Whatever its origin it provides further information on the flora of the Lancaster region and if the collector is correctly identified provides further evidence of the influence of the members of the Society. A photograph of William Hall as Mayor of Lancaster is contained in the Lantern Images Collection of Lancashire County Council.

So far in the story of the discovery of the North Lancashire flora the pre-eminence of Quakers has been pointed out. In contrast to many Lancashire natural history societies of the eighteenth and early nineteenth century the North Lancashire botanists and the members of the Lancaster Literary, Scientific and Natural History Society were not artisan naturalists. They were financially comfortable and even of private means owning property including small estates. They had a background as Yeoman farmers but had often accumulated significant wealth in trade or in developing manufacturing businesses. They lived a comfortable life but it all changed towards the end of the 1840s. The Crosfields had moved to Warrington and Liverpool, the Fieldings had gone to Bolton-le-Sands before returning to Lancaster (Plate 13) where H.B. Fielding died at the early age of 46 years in 1851 whilst Samuel Simpson became a priest. At this point the first early phase of botanical exploration ceased. However the Fieldings' era only ended with the death of Mrs Fielding in 1895. She was a wealthy lady and left £28,000, or £2.8m using the Retail Price Index value in 2013. Amongst the provisions in her will was a legacy of £1000 to the Trustees of the Oxford University Botanic Garden, which was to be invested and the income used in part payment of the stipend of the curator for ever.

There were artisan naturalists in North Lancashire although little is known about them, but their work may be contained in publications by William Hutton and Peter Whittle. At the end of the eighteenth century Blackpool was emerging as a coastal resort. At that time it was no more than a hamlet at the northern end of the Lytham sand hills on the Fylde coast and difficult to access from Preston or elsewhere. Undrained coastal marshes and peat bogs provided major obstacles to travellers. Nevertheless in 1789 William Hutton (1723–1815) published an account of the area (Hutton, 1789). He took a great interest in local history and topography and had settled in Birmingham but travelled widely and often wrote about places he visited including Blackpool (Gordon, 1891).

Sometime about 1830 another account, *A New Description of Blackpool*, was published (Anon., *c.* 1830), which included a list of plants found in the vicinity of Blackpool in 1793. It was possibly based on Hutton's work although he was not himself a botanist. If as seems likely Hutton did not write the botanical account the author remains unknown. A memoir of Hutton was also included. This new account was possibly written by P. Whittle (1789–1866) of Preston (Sutton, 1900). According to an account of his life (Whittle, 1837) Peter Whittle was born at Inglewhite to the north of Preston in 1789 and was educated at various grammar schools including Preston Grammar School. Here he had access to the Shepherd Library. Dr Shepherd is believed to be Richard Shepherd (1694–1761). He practised as a physician in Preston and was active in the town's politics becoming an Alderman in 1746 and Mayor in 1747–1748 and 1755–1756. He was a man of means and assembled a library of about 8500 volumes. In his will Shepherd left his library to the Corporation of Preston plus £200 the interest of which was to pay a librarian's salary and the residue of the estate of *c.* £1000 invested and the interest used to purchase books. It was originally in a house occupied by the head of Preston Grammar School located in Shepherd Street adjoining the Arkwright Arms. It was then removed to Cross Street, off Winckley Square.

After his school education Whittle was apprenticed to a bookseller in Preston. He then became interested in various scholarly pursuits but according to Sutton (1900) his work was often inaccurate. In a *New Description of Blackpool* the list of plants with several dubious records was compiled in 1793 and is therefore not Whittle's own work. Nevertheless in his account of Preston (Whittle, 1837) he refers to a prospectus for a 'Flora Prestoniensis' but this was never published and a manuscript has not been found. However it is doubtful if he would have been much more than the publisher. He also refers to botanical and natural history societies that met in local houses (Whittle, 1821). Their names suggest they were all public houses. The first botanical society was established under the auspices of Mr James Winstanley and Mr William Salt (both deceased by 1821) and met at the home of Mr T. Hope, Butchers Arms in Molyneux Square but moved to the house of Mr Walter Foss, Green Man, Lord Street. William Helme (1784–1834) (Desmond, 1994) was a member of this society and although primarily an entomologist collected plants with Mr Tomlinson, surgeon (Whittle, 1837). Helme was born in Warrington but became a cotton operative in Preston. Secord (1994a) provides an account of his correspondence and draws out the distinction between Helme, the artisan and the gentlemen to whom he wrote.

Another group with 70 members that split away from this society met at Mr Layfield's house, the Lamb and Packet, at the bottom of Friargate (Plate 14) where it was still present in 2000. Apart from these references nothing is known of the activities of these artisan groups although it is possible they are responsible for the eighteenth century plant list for Blackpool and perhaps drafted a 'Flora Prestoniensis'.

Although Peter Whittle and H.B. Fielding came from different social backgrounds they had a common influencing factor. Whittle in his account of his life says he was taught by Dr Shepherd but he died many years before Whittle was born. However it is assumed he had access to

Shepherd's library when he was a pupil at Preston Grammar School. The library was also made available to H.B. Fielding when Robert Harris (1764–1862) tutored him. He was the incumbent at St George's Church, Preston, 1792–1862 and his son, Edmund, was to become one of Preston's greatest benefactors (Hunt, 1992). The Shepherd Library was a source of enjoyment to Fielding[10] and was no doubt of great benefit to and influence on Peter Whittle.

In 1837 William Thornber (1803–1885) published *An historical and descriptive account of Blackpool and its neighbourhood* (Thornber, 1837). The area covered was clearly defined and the list of plants included seems fairly accurate. However, although Thornber was a colourful character and had a sad if not tragic later life (Clarke, 1923; Stott, 1985) he was not a botanist. In his introduction he thanks Mr Kenyon of Preston for the contribution on natural history. Unfortunately nothing is known of Mr Kenyon. Stott (1985) gives a detailed account of Thornber's life from which it is clear he came from a family of means. In his early life he took a great deal of interest in cultural affairs and was involved in the development of Blackpool. Although possibly of a similar social standing and with similar interests there is no evidence that he met H.B. Fielding and his circle.

Finally there is one record from the Preston area that is of considerable interest. In the herbarium of the Manchester Museum there is a sheet of Pillwort (*Pilulifera globulifera*) from Ribbleton Moor, Preston, collected in 1825. Unfortunately no collector is named and the Manchester Museum is not clear how or when the specimen was acquired. The specimen is notable as it is the only record for this species from Lancashire north of the R. Ribble. It also raises the possibility that it was gathered by one of the Preston artisan naturalists.

Clearly there was a lot of interest in the flora of Preston, Lytham and Blackpool in the early nineteenth century and it is likely that Whittle's account (1821) of groups meeting in Preston public houses suggests that the members were artisan naturalists for which South Lancashire and the Mersey Basin area is so well known (Percy, 1991; Secord, 1994b). No doubt they assembled collections (it is known that William Helme formed an entomological collection, donated to the town but subsequently lost) and it is a great pity that almost nothing survives. Socially they would not have met Fielding who as a boy lived in Preston and later visited Lytham on holiday. Yet despite this Dr Shepherd's library may have influenced both Fielding and the Preston artisan naturalists.

North Lancashire Botany 1850–1925

From the middle of the nineteenth century the development of railways enabled visitors and residents alike to get to hitherto unexplored parts of North Lancashire, although some areas of Bowland remained unvisited except by the most intrepid explorers. Even by 1922 when all the railways in the region were open some parts, especially in Bowland, were more than five miles from a railway (Crosby, 2006; fig.4.14).

The railway reached Preston from Wigan in 1838. Further lines were opened to Longridge, Fleetwood and Lancaster in 1840. Lines to Blackpool, Lytham, St Anne's and Knott End followed. From Lancaster routes were opened to Morecambe, later extended to Heysham and Wennington in 1848 and 1849, whilst the Furness and Midland Joint Railways opened a railway linking Wennington with Silverdale. Another line was built from Lancaster to Glasson Dock. Thus by 1900 much of North Lancashire was accessible by foot from a local railway station although walks of twenty miles were required to explore western parts of Bowland (Holt, 1978). However eastern Bowland near Slaidburn and the upper reaches of the Hodder valley remained largely inaccessible.

C.J. Ashfield was probably the first botanist to take advantage of the new transport facilities. Charles Joseph Ashfield (1818–1877) was born in Buckinghamshire and trained to become a solicitor. After a short period in Islington, London (1841 census) he moved to Preston where he died. Whilst in Preston he wrote a series of articles on the flora of the Preston area both north and south of the R. Ribble (Ashfield, 1858, 1860, 1862 and 1865). There is also a manuscript 'Flora of Preston' at World Museum, Liverpool that gives additional detail to the published records. Unfortunately this was overlooked in compiling the *Flora of North Lancashire* (Greenwood, 2012). Most of the records appear to be his own observations but he acknowledges the contribution of others especially in the later papers. Particularly notable are the observations of James Pearson (1823–1877) of Rochdale but originally from the Fylde.

Ashfield's work is notable for the first observations on the wet heath of Ribbleton Moor to the east of Preston and of plants colonising the Lancaster Canal. In addition he wrote papers on the flora of Silverdale (Ashfield, 1864) and Lytham (Ashfield, 1861). Ashfield also contributed to the floras of other areas. In 1862 he published a note on Lungwort (*Pulmonaria officinalis*) in Suffolk (Simpson, 1982) based no doubt on observations made whilst visiting relatives in the county where the Ashfield family originated. He also contributed to the *Flora of Buckinghamshire* (Druce, 1926) based on observations made when he was a boy or on visits to his parents. Ashfield's herbarium or at least part of it was acquired by Preston Scientific Society before eventually being deposited at World Museum, Liverpool.

During the second half of the nineteenth century H.C. Watson was publishing his work on the distribution of flowering plants in the British Isles (Watson, 1883). However he may have had difficulty in obtaining data from Lancashire. His herbarium at the Royal Botanic Gardens, Kew contains material gathered by Samuel Simpson (see above) but much of his information seems to have relied on a young curate at St Paul's Church, Preston in the 1870s. This was Edward Francis Linton (1848–1928). He went on to become one of the most famous British botanists of 100 years ago. However he published a list of plants growing in V.C. 60 (Linton, 1875) but annoyingly there are few localities in the list or voucher specimens in his herbarium.

Following the work of Hall and Ashfield in the 1850s and early 1860s little further exploration of North Lancashire's flora took place until the 1890s. This lean period was reflected nationally in the fortunes of the Botanical Society and Exchange Club (Allen, 1986) but, paradoxically, at the local level in some areas coincided with the publication of several county floras (Allen, 1976). These included *The Flora of West Yorkshire* (Lees, 1888). This Flora included that part of Bowland now in Lancashire but then in Yorkshire. Unfortunately it includes little about the flora of Bowland and it is not clear if Lees knew the area very well. However in an earlier work (Davis and Lees, 1878) Lees refers to one brief excursion he made with a friend to the upper Hodder valley and provided a short list of the plants known from the area. Otherwise he refers to the area as '*terra incognita*'.

Returning to the Preston area Frederick Charles King (1847–1911) wrote a chapter on the geology, botany and physical history in *A History of Longridge* (King, 1888) in which he also provided the drawings for the engravings in the book. This suggests that King was a competent botanist yet apart from the few years when he lived in Preston (*c.* 1879–1885) little is known about his botanical interests. He was born in Oxfordshire and became an Inland Revenue Officer. He met his wife in London but then moved about the country being stationed in Leicestershire, Northamptonshire and Greenock near Glasgow, Banffshire and Sussex where it is believed he died. His small herbarium containing voucher specimens for the Preston area was donated to the Natural History Museum, London from an address in Swansea long after he died.

King's work heralded a 'golden age' for the study of the North Lancashire flora. In 1891 a second edition of the *The Flora of Stonyhurst District* was published (Anon., 1891), a preliminary edition having been published in 1886 (Anon., 1886). In a preliminary note to a later edition Turner (*c.* 1986) suggests that the authors were Fathers C. Newdigate and H. Wright. Fr Newdigate became a well-known scholar but so far the life of Fr Wright has not been traced. The authors of the second edition were probably Fr Newdigate and Fr Gerard. It covered parts of Lancashire north and south of the R. Ribble as well as parts of what was then Yorkshire. These were the first records for this part of northern Lancashire and voucher specimens remain at Stonyhurst College (Plate 15).

Albert Wilson (1862–1949) was born at Garstang and according to his own account published posthumously (Wilson, 1953), he became interested in plants whilst at school in Kendal in 1876. He was encouraged by his parents to study botany and for the next thirty years he largely pursued his interest on his own. Although he was often absent from Lancashire, his parents remained at Garstang living at Calder Mount on Bruna Hill (Plate 16). Albert Wilson trained as a pharmacist and ran a business in Bradford travelling daily from his home in Ilkley. In 1898 he met James Alfred Wheldon (1862–1924) also a pharmacist (Wheldon, 2011). He was already a well-known botanist with a particular interest in bryophytes. This became a close friendship until Wheldon died in 1924. However by the time Wilson met Wheldon the former had already amassed many specimens, notes and records from V.C. 60, West Lancaster. Today this vice-county has little meaning to most people but encompassed Lancashire north of the R. Ribble including the parish of Dalton now in Cumbria but excluding Furness, and that part of the present county then in Yorkshire (eastern Bowland). With this knowledge and numerous local contacts Wilson suggested to Wheldon that they should write a Flora of West Lancashire (i.e. V.C. 60 and not the current Borough of West Lancashire that is south of the R. Ribble in V.C. 59). Wheldon readily agreed and for the next nine years they devoted their spare time to exploring the vice-county assembling data for the Flora. They were also able to draw on the expertise of nationally known botanists. Preparatory to the Flora they published a number of papers that provide additional details to that published in the Flora (Wheldon and Wilson, 1907). The Flora is a model of its kind with the introductory chapters adding meaning and understanding to the systematic text. In particular the topographical account highlighting areas of special botanical interest was of immense help in the preparation of the *Flora of North Lancashire* (Greenwood, 2012).

The people who provided information for Wheldon and Wilson's Flora are acknowledged by them and further details are included in the present biographical index (see below). Additionally, a number of other projects were taking place concurrently.

In 1902 the local school teacher, John Moss (1853–1925) and minister, Phipps J. Hornby (1853–1936) at St Michael's-on-Wyre ran a project whereby the pupils brought to school each week plants collected locally. John Moss often supplemented these records with plants he had found, sometimes further afield in Lancashire. Some years later, probably based on these records, Moss and Hornby compiled a Flora of the Parish (Hornby and Moss, 1925). Unfortunately little detail was given in the publication but the original manuscript and notes survived (now in the possession of Lancashire Museum Service). This gives a great deal more information providing details of the farms where the plants were found and the names of the pupils who found them. Only some of this data was published by Wheldon and Wilson (1907). The barrister, Hugh Phipps Hornby (1849–1944) was a brother of Phipps J. Hornby and amateur ornithologist (Oakes, 1953). They were members of a well-known Fylde family

Preston Scientific Society was founded in 1876 but failed after four years, reforming in 1893. An early project of the botanical section of the Society was to publish a *Flora of Preston and Neighbourhood*. This was based on records collected between 1897 and 1902 and published in 1903 (Preston Scientific Society, 1903). Those contributing to the Flora are acknowledged and further details about most of them are provided in the biographical index (see below). However it appears that the lead was taken by William Clitheroe (1864–1944). He was a Preston school teacher but retired to Bowness-on-Windermere, where he died in 1944 aged 80.

Wheldon and Wilson acknowledged the contributions made in the Silverdale area by several members of the Pickard family. Joseph Pickard was born in Silverdale in 1876 and was a Leeds draper. His botanical interest was probably encouraged by his family as he took an early interest in field botany and published a note of the flora of Newton-in-Bowland in the *Natural History Journal* aged seventeen (Pickard, 1893). This was the start of many years of botanising in the area with papers published in *The Naturalist* and voucher specimens contained in his herbarium now at Leeds University. His base was The Heaning at Newton-in-Bowland where he spent several summer holidays. From here he was able to take long walks into the upper Hodder valley, the Trough of Bowland and the surrounding fells. His accounts are the first substantial and localised records of the flora of the region and describe the area before Stocks Reservoir was built and Gisburn Forest planted in the 1920s. Amongst his collaborators was Miss M.N. Peel, later Mrs Nicholas Assheton. She lived at the family home at Knowlmere, Newton-in-Bowland (Plate 17) and published two papers in *The Naturalist* (Peel, 1913a & b). The Peel family originally came from the Craven area of Yorkshire but one branch lived at Bolton-by-Bowland. There are numerous descendants but the Peels of Knowlmere and of Church including Sir Robert Peel (1788–1850) are all related (A. Peel, 1913).

Wheldon and Wilson also acknowledged the work of Samuel Lister Petty (*c.* 1860–1919) who was a man of private means. He lived in Ulverston but wrote two papers on the flora of Leck and Silverdale in *The Naturalist* (Petty, 1893, 1902).

A manuscript 'Flora of Preston' was written by Arthur Augustine Dallman (1883–1963) in 1901 when he was a teenager living in Preston. This describes his discoveries in and around Preston, especially in Lea and Fulwood. In addition to detailed descriptions of localities he also provides sketches of localities. A copy of his manuscript is at World Museum, Liverpool.

In addition to those who published papers etc. on the North Lancashire flora many others contributed records. Thus Thomas Greenlees, whose herbarium is at Bolton Museum, explored the Brock valley. Others came on holiday, e.g. W.P. Hiern from Exeter (herbarium at Exeter Museum) or F.A. Lees from Leeds who published notes on the flora of Bare near Morecambe (Lees, 1899).

The period around 1900 was one of considerable botanical activity with most of northern Lancashire accessible by rail. Nevertheless, exploring the upper Brock valley and the Bleasdale Fells involved long walks from the station at Brock. In many cases overnight accommodation was required (Wilson, 1953). It is also clear from the records and descriptions of their walks that access to private estates whether in the fells or in river valleys did not pose a problem.

The final years of this period after the *Flora of West Lancashire* was published saw little further field work. An exception was Charles Bailey (1838–1924) whose move to the developing resort of St Anne's was made possible by a fast train service to Manchester where he had his business. In the few years he was at St Anne's he made numerous collections of the introduced plants colonising the sand dunes before they were built on (Plate 18). Some of these were included at the back of the *Flora of West Lancashire* (Wheldon and Wilson, 1907) but full accounts were published in *Manchester Memoirs* (Bailey, 1902, 1907a, 1910). He also made a special study of Large-flowered Evening-primrose (*Oenothera glazioviana*) on the sand dunes (Bailey, 1907b).

For a few years around 1915 the Rev. Alfred John Campbell (1858–1931) lived in St Anne's. He was a Primitive Methodist Minister and made a small collection of V.C 60 plants that he donated to the library at St Anne's.

The end of this period culminated in a short paper noting additions and extinctions (Wheldon and Wilson, 1925). These years were dominated by the First World War, which inevitably had an adverse effect on people's leisure activities.

The Middle Years of the twentieth Century (1926–1963)

The period from the First World War until the publications of the *Atlas of the British Flora* in 1962 (Perring and Walters, 1962) was a lean period for exploring North Lancashire. A few notes appeared in *The North Western Naturalist*, notably a paper by R.S. France of Garstang (France, 1931a) and in the publications of the Botanical Society of the British Isles, mostly by H.E. Bunker and J.A. Whellan.

In 1931 R.S. France published two papers (France, 1931a & b) on the flora of West Lancashire and formed a herbarium parts of which are at World Museum, Liverpool. However his interests developed elsewhere and he became the first Lancashire County Archivist in 1940 (Plate 19).

Also in the 1930s Winifrid Gibbs, a teacher at the then Lark Hill Convent School in Winckley Square, Preston wrote a manuscript 'Flora of Preston'. Two copies dated 1939 are deposited in the Lancashire Record Office and at World Museum, Liverpool. Unfortunately they were not known to the author when the *Flora of North Lancashire* was written (Greenwood, 2012). The manuscript was more of an ecological account rather than of a systematic text. Nevertheless it provides a good description of some of the species and habitats still surviving at that time.

H.E. Bunker and J.A. Whellan collaborated in exploring the Lancashire flora although Whellan was some fifteen years younger. Both contributed records to the Botanical Society and Exchange Club of the British Isles but only Whellan published any papers or notes (Whellan, 1942, 1954). Bunker retired as the manager of the Leyland Rubber Company but in later years did little field botany due to a heart condition. He formed an herbarium but after his death it suffered insect damage and was in a poor condition when several years later it was donated to World Museum, Liverpool. Whellan was primarily an entomologist and went on to have a distinguished career in pest control in tropical parts of former British colonies before emigrating to Australia where he died in 1995 (PED, 1996).

The last half century (1963–2011) – Personal involvement

The publication in 1962 of the *Atlas of the British Flora* (Perring and Walters, 1962) was a ground-breaking book heralding a new way of recording the distribution of plants (and animals). Combined with advances in information technology, taxonomy, ecology and biogeography considerable advances were made to our understanding of the living world and the environment following its publication (Braithwaite and Walker, 2012; Preston, 2013).

For me the impact was to govern my future life. However the immediate result was the discovery that remarkably little was known about the North Lancashire Flora and that records in the *Atlas of the British Flora* relied heavily on Wheldon and Wilson's *Flora* of 1907. Consequently many common species were unrecorded.

This realisation led members of the Natural History Section of Preston Scientific Society to look through their diaries and compile a list of additions for species not recorded in the *Atlas*. The lack of recent records was confirmed by the late Drs John Dony and Franklyn Perring (pers. comm.) and it was suggested that I become the Botanical Society of the British Isles' Recorder for V.C. 60, West Lancaster shortly after the Society had established these posts nationwide. I held the position until 2013. The view was that I might prepare a supplement to Wheldon and Wilson's Flora of 1907 but in doing so the records should be collected on a 2 x 2km square (tetrad) basis.

Accordingly through the auspices of Preston Scientific Society it was decided to initiate a survey of the vice-county in 1964 by allocating 10 x 10km squares (hectads) to individual members of the Society and others. An early allocation of tetrads was as follows:

SD32 Miss A.E. Ratcliffe
SD33 & 34 Miss E.J. Harling except tetrads 33N & P, 34K, which were allocated to Mrs E.M. Pearce
SD43 & 44 Mrs E. Hodgson and J. Hodgson
SD46 Dr Joan Wilson
SD47 H. Shorrock and later Mrs C. Jones-Parry
SD53 & 54 E.F. Greenwood except tetrad 53X which was allocated to Miss N. Carbis
SD55B Miss W. Gibbs
SD56 Miss I Hodson
SD57 Prof. C.D. Pigott
SD63 Blackburn Naturalists' Field Club, J. Ainsworth
SD64 B. Oddie
SD66 W. Fiddler
SD67 tetrads A & B Miss D. North of The Heaning, Newton-in-Bowland

In 1964 I was teaching in Reading but my base remained the family home in Preston. However it proved difficult to allocate all the 10 x 10km squares to recorders and some remained unallocated or few records were collected. Although records were received from all those in the allocations shown above some recorders soon dropped out for one reason or another. In a few cases contact was lost. Others more than made up for these difficulties.

Mrs Eileen Hodgson moved with her family from Reading to Preston at about this time and brought with her tremendous enthusiasm and expertise. With her son John, later to become an academic botanist at Sheffield University, they made a huge contribution to gathering records in the early years. In addition they often recorded well beyond their allocated hectads, mainly to the north and west of Preston. Their recording included the built areas of Preston where few had ever botanised. They often went out with Alice Ratcliffe who assiduously recorded in SD32. Alice was a member of the Wild Flower Society and contributed greatly to our understanding of introductions to V.C. 60. Also on the Fylde coast Jane Harling made a considerable contribution as did Mrs Edythe Pearce, often with the assistance of the Rev. C.E. Shaw of Oldham, in the Poulton-le-Fylde area.

Further afield notable contributions were made by Bernard Oddie in SD64 and Bill Fiddler in SD66 who found an unusual locality for Bird's-nest Orchid (*Neottia nidus-avis*) in Roeburndale whilst Mrs Jones-Parry made a big contribution to recording the flora of the Silverdale area.

In 1966 I returned north to a post in charge of the botany collections at the then City of Liverpool Museum, where I remained until retirement in 1998, then as Keeper of the Liverpool, now World Museum, Liverpool. This gave me a firm botanical base in the region and for ten years or so recording proceeded with enthusiasm. At this time all record keeping was manually based.

However by the mid-1970s raising a family, living in Wirral outside the vice-county and with increased work-load and responsibilities at the Museum meant I could devote little time to recording the West Lancaster flora. Also by this time some of the early volunteers were moving away or getting infirm. Nevertheless by the early 1970s it was clear that the changes since 1907 were so great that only a new Flora would do justice to the area. Also with boundary changes taking place in 1974 a large part of V.C. 64, Mid-west Yorkshire covering eastern Bowland was added to the administrative county of Lancashire. As a consequence only a Flora covering the whole of northern Lancashire would suffice but I was not in a position to do much field work in the area at this time. This meant that the wider project would have to go on hold until such time I or someone else could undertake the necessary field work.

The Lancashire Wildlife Trust was established in 1962 and from 1963 I took an active part in its affairs being a member of various committees and from time to time served on its council. Through the Trust I was able to maintain contacts, which was particularly valuable from the 1970s until my retirement in 1998. From the 1970s the North Region of the Trust based in Lancaster was particularly active. Jennifer Newton was undertaking botanical survey work in the region and following his retirement Len Livermore with his wife Pat undertook a comprehensive survey on a tetrad basis of Lancaster District. In addition they undertook specific surveys of a number of environmental features. Their work culminated in a series of publications between 1987 and 1992. Then, in the 1990s Phyllis (Phyl) Abbott undertook survey work in eastern Bowland preparatory to her *Plant Atlas of Mid-west Yorkshire* (Abbott, 2005). In addition a major contribution was made by the Steedens, father and son, who took a particular interest in the Fylde flora. Thus, although I personally was unable to gather much data between the late 1970s and mid 1990s many thousands of records were accumulated by others.

After 1998 I was able to resume an active recording role in northern Lancashire and visited most tetrads in the area during the next twelve years or so. Although living in Wirral the motorway network enabled most parts of the region to be searched within a two hour drive. During nearly 50 years of recording many changes have taken place. Specific surveys of ponds, canals and roadside verges were repeated and the changes reported in papers (Greenwood, 2003, 2005) and in the *Flora of North Lancashire* (Greenwood, 2012). Over the years herbaria, libraries and record offices were visited in order to obtain as complete a picture of the environment and flora as possible. Inevitably some sources were missed.

Eventually, over a four year period, and with the help of the Lancashire Wildlife Trust, various sponsors including a major financial contribution I received from my late brother, Dr D.J. Greenwood CBE, FRS, I was able to publish a *Flora of North Lancashire* in 2012 (Plate 20).

The years when the flora was recorded

Britain is well-known for its amateur naturalists. Their hobby of exploring the countryside and recording what they see has built up an incomparable data-base of the country's wildlife. As ideas and technologies developed their observations have become ever more detailed. Their story is as much a social history as a story of scientific endeavour and has been described by Allen (1976) and for plants in more detail by Allen's later book, *The Botanists* (1986).

Activity and interest in pursuing a hobby in natural history tended to have peaks and troughs in popularity as well as subject matter. In North Lancashire botanical peaks and troughs probably had more to do with the economy and particularly two world wars rather than with the fortunes of national societies or interests.

Figure 1 show the decades in which recorders of the North Lancashire flora were born. This demonstrates that a peak of interest was shown by people born in the middle of the nineteenth century and whose recording effort was around 1900 before the devastating effects of the First World War became apparent and which lasted until the Second World War.

This peak of interest is slightly later than that recorded by the number of people collecting and exchanging specimens as part of the botanical exchange networks that existed in Britain and Ireland in the nineteenth and twentieth centuries (Groom *et al.*, 2014).However people born in the first decades of the twentieth century became active in retirement in the 1960s. This coincided with a recruitment drive of the Botanical Society of the British Isles (Allen, 1986) and the publication of the 'Atlas' (Perring and Walters, 1962). The publication of the 'Atlas' provided the impetus for the publication of many local and county floras including that for North Lancashire (Preston, 2013). It heralded a new, more detailed and systematic methodology for recording both plants and animals. However it remains to be seen whether or not this and its successor (Preston, *et al.*, 2002) will be reflected by more amateur naturalists getting involved in recording plants. Nevertheless with the capacity to analyse and manipulate digitised records exciting times lie ahead for the interpretation of the expanding data-base.

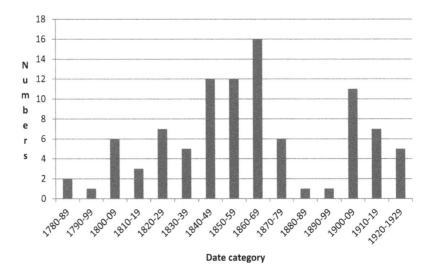

Figure 1. Birth date categories of deceased botanists

Biographical index of
Deceased North Lancashire botanists

This index gives brief biographical details of deceased botanists who have contributed to the knowledge of the North Lancashire flora. It includes botanists who have made more than a few records or have written notes about the flora but who had died before the end of 2013.

Details about some people have been difficult to find. For these only their names are recorded with an indication of their contribution. Some of the individuals date to the early nineteenth century, whilst others made notable contributions in the 1960s and early 1970s.

Where botanists have made only a small contribution to recording in North Lancashire or their lives are well-documented elsewhere reference is made to Desmond (1994) provided they are included in that work. Many dates of life events were obtained from on-line resources. Where dates refer to a year's quarter the information was obtained from the National Index to births, marriages and deaths of the General Registrar's Office. Obtaining the appropriate certificate will provide further details of their life events.

Ainsworth, John 1913–1999

Born Blackburn, Lancashire 13 June 1913, died Blackburn February 1999.

John Ainsworth had a confectioner's shop in Bolton Road, Blackburn but lived in Mosley Street. He is remembered as one of the best naturalists in the area in the second half of the twentieth century. He was involved for many years with several societies, e.g. the British Naturalists' Field Club and the North East Naturalists' Union. He was secretary of the Blackburn Naturalists' Society and for a few years in the 1960s he and members of the society helped collect records in the hectad SD63.

Ashfield, Charles Joseph 1818–1877

Born Hartwell, Buckinghamshire 29 May 1818, died Preston, Lancashire 3rd Q. 1877.

Ashfield contributed records to local floras in England but his most important work was compiling a 'Flora of Preston' (Ashfield, 1858, 1860, 1862, 1865) whilst he was in Preston. He also wrote notes on the Silverdale flora (Ashfield, 1864) and Lytham (Ashfield, 1861). Further afield he provided records for the Flora of Suffolk (Simpson, 1982) and Buckinghamshire (Druce, 1926). Ashfield was born in Buckinghamshire at Hartwell and no doubt collected records on visits to his parents. Similarly he probably had relatives in Suffolk where the Ashfield family originated. As an adult he trained as a solicitor and after a brief time in Islington in London spent the rest of his life practising in Preston. Notes in Desmond (1994) are misleading and in particular he was never a teacher.

Atkinson, Harold Waring 1868–1946

Born Bromley, Kent 3rd Q. 1868, died Watford, Hertfordshire 4th Q. 1946.

In 1901 Harold Atkinson published a list of *Rossall Fauna and Flora* (Atkinson, 1901). This seems to be the same list as was used by Watts (1928). Unfortunately it is inaccurate and the area covered is not clear. Atkinson was a teacher at Rossall School (1901 census) but in 1911 was living in Watford as a schools examiner, author and inventor.

Baecker, Dr Margaret 1902–2008

Born Vienna, Austria 1902, died Arnside, Cumbria 25 July 2008.

Dr Margaret Baecker was born in Vienna and came to the UK with her husband before the Second World War to escape Nazi persecution. She came to Arnside in 1970 and became a member of the Arnside and District Natural History Society and was President in 2005. In due course she took charge of their botanical recording scheme (1972–1987) and compiled detailed records for both Cumbria and Lancashire. These were added to a computer data-base at a later date by Charles Bromley-Webb (see below) and the records incorporated in to the North Lancashire Flora data-base. However Dr Baecker made a particular contribution with her records of *Cotoneaster* spp, and *Sorbus* spp. (with Dr Tim Rich; Rich and Baecker, 1986) in the Silverdale area. An appreciation of her long life is given by Foley (2008).

Bailey, Charles 1838–1924

Born Atherstone, Warwickshire 14 June 1838, died Torquay, Devon 14 September 1924.

Charles Bailey was a Manchester businessman but for a few years in the early 1900s lived at St Anne's on the Fylde coast. At that time development of the Clifton Estate was turning the township into a seaside resort. In the few years that Bailey lived in the town he made a study of *Oenothera* spp. (Bailey, 1907b) and of the introductions he found on the sand dunes (Bailey, 1902; 1907a; 1910). Voucher specimens for many species are at the Manchester Museum and elsewhere. The label data is particularly detailed so that precise locations in what is now the built up area of the resort can be identified. He was a member of the Botanical Society and Exchange Club and was its President, 1879–1903. Many of his collections were distributed by the Society at a time when he organised the exchange scheme. For further information see Allen (1986) and Desmond (1994).

Barbour, Henry *c.* 1837–?

Born Macclesfield, Cheshire *c.* 1837, died ?

Henry Barbour was a member of Preston Scientific Society and contributed to their *Flora of Preston* (Preston Scientific Society, 1903). He appears in the 1891 census aged 54 as a provender dealer and seed merchant in Preston but no other information is known. However his birth was registered in the 3rd quarter of 1840. His death has not been traced.

Beanland, Joseph 1857–1932

Born Bradford, Yorkshire 1857, died Bradford 18 September 1932.

Joseph Beanland lived in Bradford where he was President of Bradford Naturalists. He contributed a few records to the *Flora of West Lancashire* (Wheldon and Wilson, 1907). For further information see Desmond (1994).

Beesley, Henry 1855–1925

Born Preston, Lancashire 3rd Q. 1855, died Preston 3rd Q. 1925.

Henry Beesley was a railway clerk and lived all his life in Preston. He contributed to both the *Flora of Preston* (Preston Scientific Society, 1903) and the *Flora of West Lancashire* (Wheldon and Wilson, 1907). For further information see Desmond (1994).

Bennett, Arthur 1843–1929

Born Croydon, Surrey 19 June 1843, died Croydon May 1929; FLS 1881.

Arthur Bennett was a well-known British botanist who contributed a few North Lancashire records in a supplement to H.C. Watson's *Topographical Botany* in 1905. See Desmond (1994) for further information.

Boswell, Dr John Thomas Irvine (later Boswell-Syme) 1822–1888

Born Edinburgh 1 December 1822, died Balmuto, Fife 31 January 1888; FLS 1854.

Boswell was a well-known British botanist who contributed a few North Lancashire records to H.C. Watson's *Topographical Botany*. For further details see Desmond (1994).

Breakell, Arthur 1865–1934

Born Woodplumpton near Preston, Lancashire 1865, died Garstang, Lancashire 4th Q. 1934.

Arthur Breakell was a veterinary surgeon who lived at Barnacre-with-Bonds near Garstang. He contributed records to the *Flora of West Lancashire* (Wheldon and Wilson, 1907).

Bromley-Webb, Charles 1927–2011

Born Portsmouth, Hampshire 4 December 1927, died Arnside, Cumbria 11 April 2011.

Charles Bromley-Webb (Plate 21) was a chemical engineer and after a career with UKAEA and British Nuclear Fuels, retired to Arnside in 1984.

Here he became a member of the Arnside and District Natural History Society and succeeded Margaret Baecker (see above) as their botanical recorder. Margaret had compiled a card index of the society's records and Charles undertook the task of converting these to a computerised data-base. The records then became more widely available and the Lancashire records were incorporated into the data-base for the *Flora of North Lancashire*. Charles was a keen naturalist and visited many parts of the country as well as contributing a large number of records locally. His enthusiasm and good humour for looking at flowers was infectious and was enjoyed by the walking and natural history groups he walked with.

Bruxner, Christopher J. 1927–2009

Born Kensington, London 16 September 1927, died Kings Lynn, Norfolk 21 April 2009.

Christopher Bruxner was a land surveyor but it was whilst he was undertaking an MSc course at the Lancashire College of Agriculture and Horticulture at Myerscough Hall that he undertook a botanical survey (with Simon Hilton) of Myerscough Hall and Lodge Farms in 1986[16]. This led to a lasting interest in botany and when he was living at Pilling contributed many records to the *Flora of North Lancashire*. Later he moved to Kings Lynn where he died in 2009.

Buckley, J. *fl. c.* 1840

Nothing is known about J. Buckley other than the publication of a list of plants from Lytham in 1841 (Buckley, 1842). An address in Fitzroy Square in London was given but there is no trace of him in the 1841 census.

Bunker, Herbert Edwin 1899–1969

Born Lewisham (Catford), London 11 August 1899, died Preston, Lancashire 16 January 1969.

H.E. Bunker was a well-known Preston botanist in the 1940s and 1950s. He contributed records to the Botanical Society and Exchange Club and its successor, the Botanical Society of the British Isles, sometimes in collaboration with W.A. Whellan (see below) as well as to *Travis's Flora of South Lancashire* (Savidge, *et al.*, 1963). He was employed as a manager by the Leyland Rubber Company but in later years he suffered from heart disease that prevented him doing any field work for the *Flora of North Lancashire*. After his death his herbarium suffered from insect damage but the remains are at World Museum, Liverpool. An obituary was published in the *Lancashire Evening Post* for 17 January 1969. See also Desmond (1994).

Burns, J, *fl. C.* 1950

J. Burns wrote an article on plants near Lancaster in the *New Biologian* in 1955 (Burns, 1955) but nothing is known of his life.

Campbell, Rev. Alfred John 1858–1931

Born Shotley Bridge, Co. Durham 3rd Q. 1858, died Lanchester, Co. Durham 11 August 1931; FLS 1889.

Rev. Campbell was a Primitive Methodist Minister and had a life-long interest in botany. Whilst at St Anne's (*c.* 1915) he compiled a small herbarium of North Lancashire plants that he donated to the local library. Other collections from elsewhere are at National Museum of Wales and Hancock Museum, Newcastle-upon-Tyne. Bowran (1932) published a short obituary and Graham (1988) outlined his contribution to Durham botany.

Carbis, Nellie 1904–1999

Born Newton-le-Willows, Lancashire 23 April 1904, died Preston, Lancashire 29 November 1999.

Nellie Carbis was born and brought up in Newton-le-Willows and from an early age was a lover of plants and gardening. She was trained as a teacher and came to Grimsargh near Preston as head of the Parochial School in 1935. On retirement in 1964 she remained in the village until her death in 1999. In 1942 she acquired a garden for the school and when she could no longer look after it Grimsargh Parish Council acquired it and after renewed maintenance 'The Old School Garden' was re-named the 'Nellie Carbis Millennium Woodland' and officially opened in 2000. Nellie was a long standing member of the Natural History section of Preston Scientific Society and contributed records to the *Flora of North Lancashire* especially from tetrad SD53X. She was a great character and story teller. She published her autobiography in 1978 (Carbis, 1978).

Carr, Amos *c.* 1829–1884

Born Frant, Sussex *c.* 1829, died Sheffield, Yorkshire 29 April 1884.

Amos Carr was primarily a South Yorkshire botanist (Coles, 2011) but contributed to the *Flora of West Lancashire* (Wheldon and Wilson, 1907). He was a postman and shoemaker.

Carter, James Gunson 1862–1903

Born Colton, near Ulverston, Cumbria 4th Q. 1862, died Preston, Lancashire 1st Q. 1903.

James Carter contributed to the *Flora of Preston & neighbourhood* (Preston Scientific Society, 1903). He was a jobbing printer and compositor.

Clitheroe, William 1864–1944

Born Preston, Lancashire 3rd Q. 1864, died Bowness-on-Windermere, Cumbria 21 May 1944; FLS 1903.

William Clitheroe was chairman of the botanical section of Preston Scientific Society and contributed to the *Flora of Preston & neighbourhood* (Preston Scientific Society, 1903) and the *Flora of West Lancashire* (Wheldon and Wilson, 1907). He also checked the list of plants published by Heathcote (1923). He was a school teacher.

Crombie, Rev. William *c.* 1842–1912

Born Scotland *c.* 1842, died Burnley, Lancashire 4th Q. 1912.

William Crombie was a Congregational Minister and whilst living at Newton-in-Bowland corresponded with Joseph Pickard (see below) and contributed several records for his papers.

Crosfield, George 1754–1820

Born Kendal, Cumbria 22 March 1754, died Lancaster 10 October 1820.

George Crosfield contributed some of the earliest records to the Lancashire flora. He was a member of a well-known Quaker family with a grocer's business in Warrington and a sugar refinery in Lancaster. He and other members of the family probably influenced Henry Borron Fielding and his wife Mary Maria (see below) to take up botany and natural history as a life-long hobby. Crosfield formed an herbarium some of which is still extant (Salmon, 1912). See also Desmond (1994).

Crosfield, George 1785–1847

Born Warrington, Lancashire 26 May 1785, died Liverpool 15 December 1847.

George Crosfield (1785–1847) followed his father's botanical interests collecting plants in the Lancaster area. He also followed his father in his business interest and spent most time in Warrington. It was here that he did most of his botanical work. A few of his herbarium specimens are extant (Salmon, 1912). See also Desmond (1994).

Cross, William *fl.* 1890

William Cross wrote an article on Fylde flowers (Cross, 1889) and made one of the earliest records for an evening primrose from the sand dunes at Lytham St Anne's. Unfortunately the species is not identified. The identity of William Cross has not been ascertained.

Crump, William Bunting 1868–1950

Born Scarborough, Yorkshire 26 April 1868, died Wharfedale, Yorkshire 1st Q. 1950.

William Crump was a Yorkshire botanist and science master who contributed records to Wheldon and Wilson (1925). See also Desmond (1994).

Cryer, John 1860–1926

Born Bailden, Yorkshire 29 July 1860, died Shipley, Yorkshire 7 May 1926.

John Cryer was a Yorkshire botanist and teacher. He contributed several records from the Silverdale area to Wheldon and Wilson (1925). He was especially interested in hawkweeds. See also Desmond (1994).

Curwen, Edwin 1858–1920

Born Barnacre-with-Bonds, Lancashire 1st Q. 1858, died Garstang, Lancashire 3rd Q. 1920.

Edwin Curwen was a stonemason and contributed records to the *Flora of West Lancashire* (Wheldon and Wilson, 1907).

Dallman, Arthur Augustine 1883–1963

Born Stannington, Northumberland 9 April 1883, died Colwyn Bay, Denbighshire 20 March 1963; ALS *honoris causa* 1935.

Arthur Dallman was the son of a Church of England Minister and moved to Preston at an early age. When only 17 he wrote a MS Flora of Preston (now at World Museum, Liverpool) in 1901. This provides detailed references to plant localities in the Ashton and Lea areas of the town. At that time he was a chemist's apprentice but by 1901 he had moved to Liverpool where he was a teacher. Subsequently he moved to various teaching posts in Britain amongst which was a post at Mexborough Grammar School, Yorkshire where he remained for 20 years. In 1954 he retired to Colwyn Bay where he died. He is best known for founding and editing the *North Western Naturalist* but wherever he went he recorded the flora. However his MS Floras of Flint and Denbighshire (now at World Museum, Liverpool) were not published. Further details can be found in Greenwood (1977), Coles (2011) and Desmond (1994).

Dillwyn, Lewis Weston 1778–1855

Born Ipswich, Suffolk 21 August 1778, died Swansea, Glamorgan 31 August 1855; FRS 1804, FLS 1800.

Lewis Weston Dillwyn is best known for *Botanist's Guide through England and Wales*, which he published with Dawson Turner (Dillwyn and Turner, 1805). This contains several early records for North Lancashire. See also Desmond (1994).

Dobson, William 1820–1884

Born Preston, Lancashire 1820, died Chester 8 August 1884.

William Dobson was a Preston journalist and stationer. He is best known for his *Rambles by the R. Ribble* where, in the third series (Dobson, 1883), he makes particular reference to the wet heath flora of Ribbleton Moor. An obituary was published in the *Preston Chronicle and Advertiser* for 16 August 1884. See also Desmond (1994).

Dunlop, Margaret *fl.* 1930s

Margaret Dunlop wrote a paper on the phytogeography of the Fylde (Dunlop, 1936) but nothing has been found about her. Her paper seems to have relied on the work of Wheldon and Wilson (1907).

Fielding, Henry Borron 1805–1851

Born Garstang, Lancashire 17 January 1805, died Lancaster 21 November 1851; FLS 1838.

Fielding, Mary Maria (née Simpson) 1804–1895

Born Lancaster 25 January 1804, died Lancaster 22 February 1895; wife of H.B. Fielding, married 21 December 1826; sister of Samuel Simpson, see below.

H.B. Fielding devoted his life to the study of botany but in Lancashire his contribution to the knowledge of the flora is limited to the comments attached to the paintings his wife made of plants found in the Lancaster area. He formed a large herbarium of plants he acquired from overseas collectors. After his death his collection passed to Oxford University, which his wife later endowed with a legacy. He had started to train as a solicitor in Liverpool where his future wife was training as an artist. After marriage they lived for many years at Stodday but they were sufficiently wealthy not to seek employment. See also Desmond (1994), which has references to several obituaries. An obituary of Mrs Fielding was published in *Lancaster fifty years ago January 1852–December 1853* extracted from the *Lancaster Guardian* and published by E. & J.L. Milner, Lancaster. Copy at Lancaster City Library.

Fiddler, William (Bill) 1901–1972

Born Preston, Lancashire 20 April 1901, died Preston 1st Q. 1972.

Bill Fiddler was a member of the natural history section of Preston Scientific Society and in the late 1960s recorded for the *Flora of North Lancashire*, mainly in the hectad SD66. His most notable discovery was of Bird's-nest Orchid (*Neottia nidus-avis*) in Roeburndale in 1965. He was a confectioner of Eldon Street, Preston.

Fisher, Harry 1860–1935

Born Nottingham 3 June 1860, died Grantham, Lincolnshire 21 January 1935.

Harry Fisher was a chemist in Newark and primarily a Nottinghamshire and Lincolnshire botanist but he also wrote the botanical sections to the Lancashire and Leicestershire volumes of the Victoria County Histories. The Lancashire volume (Fisher, 1906) was published shortly before the *Flora of West Lancashire* (Wheldon and Wilson, 1907) and appears to rely heavily on their work. See also Desmond (1994).

Fowler, Rev. William 1835–1912

Born Winterton, Lincolnshire 27 February 1835, died Winterton 7 March 1912.

The Rev. Fowler was vicar of Liversedge in Yorkshire and a well-known botanist. Inspired by reference to the possible scope for botanical discoveries in the Slaidburn area by Davis and Lees (1878) Fowler decided to spend a few days exploring the upper Hodder valley in 1881 and again in

1882. He published a list of his records in the *Naturalist* (Fowler, 1881), which are amongst the earliest for the region. See also Desmond (1994) and Coles (2011).

France, Reginald Sharpe 1909–1986

Born Fylde, Lancashire 12 November 1909, died Garstang, Lancashire 18 October 1986.

Reginald Sharpe France was the first County Archivist for Lancashire County Council appointed in 1940. However in the 1930s he was an active botanist. Herbarium specimens of his are at World Museum, Liverpool, whilst he published papers in the *Naturalist* and *Countryside* (France, 1931a & b) on the North Lancashire flora. Also a copy of Wheldon and Wilson (1907) at Lancaster University contains annotations, which are believed to be by him. An obituary, with portrait, was published in the *Lancashire Evening Post* on 21st October 1986

Frankland, Joseph Norman 1902–1995

Born Settle, Yorkshire 10 November 1902, died Newby near Clapham, Yorkshire May 1995.

Norman Frankland was a woodworker and all-round naturalist who lived at Giggleswick near Settle. He formed a fine herbarium now at World Museum, Liverpool and *A Flora of Craven* by him was published after his death (Frankland, 2001). This covers a small portion of North Lancashire. In North Lancashire he also botanised more widely in the Lune valley and his records were used in the compilation of the *Flora of North Lancashire* (Greenwood, 2012).

Fraser, Dr John 1820–1909

Born Glasgow 22 March 1820, died Wolverhampton, Staffordshire 13 April 1909.

John Fraser obtained an MD from Glasgow University in 1852 and moved to Wolverhampton in 1854. He collected plants in Staffordshire (Edees, 1972) but his herbarium is in the Geography Department of Hull University. In this there are a few plants from V.C. 60 where he seems to have visited Fleetwood in 1876 and 1882. Further details are in Desmond (1994).

Garlick, George Wade 1915–1999

Born Bentham, Settle, Yorkshire 2 August 1915, died in South Gloucestershire on 6 January 1999.

It is believed George Garlick (Plate 22) was born in Bentham near Settle and contributed an article on the Heysham salt marsh flora in the *New Biologian* (Garlick, 1957) based on observations made in 1955. He also contributed records to the *Proceedings of the Botanical Society of the British Isles* in 1957 with records from the Lancaster area covering the period 1948–1955. However during the 1950s he was also collecting records from the Bristol area and many of these were used in the *Atlas of the British Flora* (Perring and Walters, 1962). He was a school teacher and at some time in the 1950s he moved to Yate, near Bristol, where he remained for the rest of his life. He made a special study of the Avon Gorge with records from 1951 and in doing so realised the importance of the white-beams. His work is recognised in Rich *et al.* (2010) and his collections were deposited at Bristol City Museum.

During the 1950s he became interested in bryophytes and a substantial collection of nearly 6000 packets are deposited at the National Museum of Wales. Gatherings made in the 1950s are from North Lancashire and nearby Yorkshire but most are from S.W. England and Wales. He made several new vice-county records. He contributed records to Hill, *et al.* (1991–4) and Bosanquet (2003) acknowledges his special contribution to recording Monmouthshire bryophytes in the 1970s and early 1980s.

He was a member of the Botanical Society of the British Isles and the British Bryological Society between 1955 and 1965.

Gerard, John 1545–1612

Born Nantwich, Cheshire 1545, died London February 1612.

John Gerard was one of the most famous of early British botanists. He published his '*Herball*' in 1597. In an edition of 1636 amended by T. Johnson there is a record of Cloudberry (*Rubus chamaemorus*) from 'Greygarth a high fell on the edge of Lancashire...'. For further details see Desmond (1994).

Gerard, Fr John 1840–1912

Born Edinburgh 30 May 1840, died London 13 December 1912; FLS 1900.

John Gerard was a priest and teacher at Stonyhurst College. With his colleague, Charles Newdigate, he wrote, according to both Wheldon and Wilson (1907) and Desmond (1994), the *Flora of Stonyhurst* published anonymously in 1891. This was a new edition of an earlier work in which the contribution of Fr H. Wright is acknowledged. The *Flora* covered the then Yorkshire side of the R. Hodder as well as the Lancashire side north and south of the R. Ribble. The work contains some of the earliest records for that part of the region, especially for the Yorkshire bank of the R. Hodder. After leaving Stonyhurst Fr Gerard went on to lead a distinguished scholarly life. Obituaries were published in the *Journal of Botany* and the Proceedings of the *Linnean Society* in 1912 and 1913 respectively. For further details see Desmond (1994).

Gibbs, Winifrid Mary 1902–1971

Born Preston 7 July 1902, died Preston, Lancashire 1st Q. 1971.

Winifrid Gibbs was a biology teacher at Lark Hill Convent School Preston. She was a member of the natural history section of Preston Scientific Society and took a particular interest in ecology. In 1939 she compiled a MS 'Flora of Preston' copies of which are deposited at World Museum, Liverpool and with the Lancashire County Museum Service.[18] During the 1960s she volunteered to collect records for the *Flora of North Lancashire* in hectad SD55.

Gordon, Vera 1918–2006

Born Liverpool 25 May 1918, died Ormskirk, Lancashire 14 September 2006.

Vera Gordon (Plate 27) lived the whole of her life in Bootle and Liverpool where she had a career as a civil servant in the Magistrates Court. She travelled widely in this country and overseas and many journeys took her to North Lancashire. She was a prominent member of the Liverpool Botanical Society, where she was Hon. Secretary for many years, and the Botanical Society of the British Isles (B.S.B.I.) and became an honorary member of both. For the B.S.B.I. she was a referee for V.C.s 59 & 60 and District Secretary for the Mersey Province at the time the first 'Atlas' (Perring and Walters, 1962) was being prepared. Subsequently she became Recorder for V.C. 59 and contributed many records to the 2nd 'Atlas' (Preston, Pearman and Dines, 2002) as well as contributing records to the *Flora of North Lancashire*. Her herbarium is at World Museum, Liverpool along with an extensive 35mm slide collection. An obituary was published by Allen (2007).

Greenlees, Thomas T. 1864–1949

Born Bolton, Lancashire 4th Q. 1864, died Bolton 2nd Q. 1949.

Thomas Greenlees of Bolton was trained as a shoemaker but had many jobs during his life. His interest in botany started at an early age and whilst his observations were mainly in the Bolton area he visited many other places. His collections form the basis of those at Bolton Museum. Amongst these are a number of specimens collected in V.C. 60. A short biography was published by Hancock (1980). See also Desmond (1994).

Hall, W. *fl.* 1862

There is a small collection at World Museum, Liverpool of plants collected in the Lancaster area in the 1860s by a W. Hall. His identity has not been confirmed as there are several W. Halls living in Lancaster at that time. However it is likely that he is Dr William Hall born about 1819 and died 8 July 1891. He followed his father as a medical practitioner in Lancaster having graduated in Edinburgh. He curated the collections of the short-lived Lancaster Literary, Scientific and Natural History Society and was therefore an acquaintance of Henry Borron Fielding, Samuel Simpson and their circle. He became a prominent citizen and a brief testimonial to his services at the Lancaster Infirmary and Dispensary was published in the *Lancaster Gazette* for 20 February 1892 as part of a report of the AGM of the Infirmary and Dispensary. He was also Mayor of Lancaster *c.* 1878 and there is a photograph of him in his mayoral robes in the Lantern Images collection of Lancashire County Council. Although Hall's collection is small it contains a number of interesting records from a period when little botanical recording was being undertaken.

Hanbury, Frederick Jansen 1851–1938

Born Stoke Newington, London 27 May 1851, died East Grinstead, Sussex 1 March 1938.

Wheldon and Wilson (1907) acknowledged his contribution of hawkweed records in V.C. 60. See Desmond (1994).

Hardy, Eric 1912–2002

Born Liverpool 11 March 1912, died Liverpool May 2002.

Eric Hardy was a journalist and natural historian but with a primary interest in birds. He published and broadcast widely in North West England but as a Liverpudlian most contributions were published or broadcast in Merseyside. Although Eric Hardy made observations himself he relied on others, especially in later years, to provide much of the information he published. Some of his published data came from North Lancashire but it is often difficult to know by whom or when the observations were made. Notes on his life and some of his articles were re-published by Bryant (2008) but many have not been checked or transcribed. A further appreciation of Hardy's life was contained in a review of Bryant's book in the *Liverpool Daily Post* for 1 December 2009. Many of Hardy's papers are at World Museum, Liverpool and Dr I.D. Wallace published an account of these ('Eric Hardy Bequest', *Merseyside Naturalists' Association Newsletter*, December 2002).

Harling, Ellen Jane 1903–1985

Born Knottingley, Yorkshire 24 July 1903, died Blackpool, Lancashire August 1985.

Jane Harling was a Blackpool primary school teacher and in the 1960s volunteered to contribute records for the *Flora of North Lancashire,* mostly from hectads SD34 & 35. She was an assiduous recorder and contributed most of the records for those hectads in the published work. In the 1970s she moved to Cumbria where she contributed to the *Flora of Cumbria* (Halliday, 1997) but increasingly arthritis restricted her activities. She joined the committee of the British Empire Naturalists' Association Fylde Branch as 'Flower Recorder' when it was formed in 1946. She continued to serve on its committee until 1948 when the Association became the Fylde Naturalists' Society. She was then its first Chairman until 1951 but remained on the committee until the late 1950s. She was also a member of the Botanical Society of the British Isles, which she joined in 1964.

Hartley, John Wilson 1866–1939

Born Lancaster 4th Q. 1866, death registered Lancaster 1st Q. 1940 although he may have died at the end of 1939.

John Hartley had a shop in Carnforth and was regarded by J.A. Wheldon as an eminent botanist (Wheldon, 2011). Wheldon and Hartley collaborated with each other, especially in recording the Isle of Man flora. Several North Lancashire records are attributed to Hartley in Wheldon and Wilson (1925).

Heathcote, William Henry 1861–1926

Born Preston, Lancashire 5 September 1861, died Longton near Preston, Lancashire 17 February 1926; FLS 1897.

William Heathcote was secretary of the botanical section of Preston Scientific Society at the time the *Flora of Preston & Neighbourhood* was published (Preston Scientific Society, 1903). He contributed to Wheldon and Wilson (1907, 1925) and updated the 'Flora of Preston' (Heathcote, 1923). He was an all-round naturalist but was particularly interested in conchology. Census records describe him as a 'master of scientific apparatus'. An obituary with photograph was published in *The North Western Naturalist* (Jackson, 1926).

Henderson, Cl. W. *fl.* 1960s

Henderson wrote a report in the *Alpine Garden Society Bulletin* (Henderson, 1962) of a 1958 field meeting based for part of the time in Silverdale. Amongst the plants recorded was Narrow-leaved Helleborine (*Cephalanthera longifolia*) but identification was based on vegetative features. The identity of Cl. W. Henderson has not been traced.

Hiern, William Phillip 1839–1925

Born Stafford 19 January 1839, died Barnstable, Devon 29 November 1925; FRS 1903, FLS 1873.

William Hiern was a Devon botanist and landowner who visited North Lancashire on a few occasions. His records together with the voucher specimens in his herbarium (at Royal Albert Memorial Museum & Art Gallery, Exeter) have contributed to the North Lancashire flora data-base. See also Desmond (1994).

Hodgson, Eileen Nancy (née Needham) 1920–1993

Born Bethnal Green, London 3rd Q. 1920 died York December 1993.

Eileen Needham (Plate 23) married Albert Hodgson in 1944 and was a prominent botanist in Reading before coming to Preston in *c.* 1962. She joined the Botanical Society of the British Isles in 1963 and the natural history section of Preston Scientific Society. She was also a member for many years of the Wild Flower Society. For a few years in the 1960s she was an assiduous recorder of the North Lancashire flora where she took responsibility for the hectads SD43 & 44 as well as recording in the Fylde and in the Preston area more generally. With her son, John, who later became an academic botanist at Sheffield University, most of the records for tetrads in these hectads in the *Flora of North Lancashire* are attributed to them. Some years later her herbarium was found in a boarding house of the Friends' School, Lancaster. This was given to Lancaster University but their herbarium is now at Tullie House Museum and Art Gallery in Carlisle. However in 1968 Eileen Hodgson remarried and moved to York where she died. Whilst in York she was associated with botanising in the area, especially Fulford Ings.

Hornby, Venerable Phipps John 1853–1936

Born Garstang, Lancashire 10 January 1853, died Garstang 4 April 1936.

Ven. Phipps Hornby was a member of a well-known Fylde family and was part of a group of naturalists who were particularly active around 1900. He was minister of St Michael's-on-Wyre parish church (Plate 29) and with local school teacher, John Moss (see below), compiled a list of plants found in the St Michael's area (Hornby and Moss, 1925) but unfortunately without localities. He also contributed records to the Botanical Society and Exchange Club and Wheldon and Wilson (1907).

Hutton, William 1723–1815

Born Derby 30 September 1723, died Birmingham 20 September 1815.

William Hutton lived in Birmingham and in about 1788 spent some time in Blackpool in its earliest days as a resort. Following his visit he wrote *A description of Blackpool in Lancashire frequented for sea bathing* in 1789. This was re-printed anonymously in 1830 but with a memoir of Hutton and including a list of plants compiled in 1793. It was published as *A new description of Blackpool in the parish of Bispham Hundred of Amounderness* possibly by Peter Whittle of Preston (see below) However it is not known who compiled the list of plants in 1793 but it was one of the first notes describing the flora of the area.

Inchbald, Peter 1816–1896

Born Doncaster, Yorkshire 23 February 1816, died Hornsea, Yorkshire 13 June 1896; FLS 1880.

Following a visit to Blackpool he published his observations a year later (Inchbald, 1865). For further details see Coles (2011) and Desmond (1994).

Jackson, William J.S. 1859–1920

Born Preston, Lancashire *c.* 1859, died Preston 2nd Q. 1920.

William Jackson was member of the botanical section of Preston Scientific Society and contributed to the *Flora of Preston & neighbourhood* (Preston Scientific Society, 1903). He was a chemist's assistant in Preston.

Jenkinson, James 1739–1808

Born Yealand Conyers, Lancashire 6 November 1739, died Yealand Redmayne 15 October 1808.

James Jenkinson was a Quaker and teacher at Yealand Conyers and unusually for a member of the Society of Friends has a memorial marking his burial place in a copse behind his house (Plate 5). He wrote a book in 1775 (Jenkinson, 1775), which included numerous records of plants in the Silverdale area thus publishing the first account of the limestone flora of the region.

Jones-Parry, Claudia Kilvington (née Cuff) 1906–1990

Born 29 May 1906, died Weston-super-Mare, Somerset April 1990.

Claudia Kilvington Cuff married Ian H. Jones-Parry in 1936 at Bucklow, Cheshire. They came to live in Silverdale following his retirement and during the late 1960s she took a special interest in recording the flora in hectad SD47. Her records form the basis of those in the *Flora of North Lancashire.* She joined the Botanical Society of the British Isles in 1966 and by the 1970s was living in Cornwall but does not appear to have contributed to the recording of the Cornish flora (Margetts and David, 1981).

Just, Professor John 1797–1852

Born Natland, Kendal, Cumbria 3 December 1797, died Bury, Lancashire 14 October 1852.

John Just was a school teacher and taught at Kirkby Lonsdale in 1817 before moving to Bury in 1832. In Bury he became a distinguished botany teacher becoming Professor of Botany at the Manchester Institution in 1848. He was also an accomplished archaeologist and linguist. Very little is known about his botanical interests when he was in Kirkby Lonsdale but there is a reference to a catalogue, apparently now lost, of the flora that he compiled whilst in the area.[13] For further details see Desmond (1994) and the *Preston Chronicle* for 23 October 1852 for an obituary.

King, Frederick Charles 1847–1911

Born Nettlebed, Henley, Oxfordshire 1st Q. 1847, died Eastbourne, Sussex 3rd Q. 1911.

F.C. King was an Inland Revenue Officer living in Preston according to the 1881 census. He was born in Oxfordshire and worked in London where he met his wife Alice and came to Preston via Leicestershire in 1880. Shortly afterwards he wrote a chapter on the Geology, Botany and Physical History of the District in *A History of Longridge and District* (Smith, 1888). The quality of the records suggests he was a competent botanist. Wheldon and Wilson (1907) acknowledge his help but details of the remainder of King's life are not clear. Nevertheless from Ham's Excise Year Books he joined the Inland Revenue in 1868 and was located at Dunbarton (1893, 1895), Eastbourne (1900) and Inveravon, Banffshire (1901, 1905) from where he was superannuated in 1910. There is some doubt as to when or where F.C. King died but there is no further listing in Ham's Excise Year Books after he was superannuated in 1910. However, a Frederick King died in Eastbourne age 65 in the third quarter of 1911, i.e. after the 1911 census. F.C. King formed a small herbarium and this was eventually donated to the Natural History Museum in London from an address in South Wales but no further details accompanied the gift. It seems that F.C. King did not live with his wife after about 1881 and that apart from his brief period in Preston took no interest in botany.

Kirkby, William 1876–?

Born Hulton, near Leeds 1876 but death not traced.

William Kirkby was a mining engineer of Leeds and contributed to Petty's note on Silverdale plants (Petty, 1902) and *The Flora of West Lancashire* (Wheldon and Wilson, 1907).

Lawson, Thomas 1630–1691

Born near Settle, Yorkshire 10 October 1630, died Great Strickland, Cumbria 12 November 1691.

Thomas Lawson was one of the early Quaker botanists. Most of his records are from Cumbria (Halliday, 1997) but he made a few records from Lancashire. Unfortunately it is difficult to be sure if his records are his own or of others. Later in life he corresponded with John Ray. For further details of his life and work see Whittaker (1986), Wilson (1744), Raven (1948), Lankester (1848) and Desmond, (1994).

Lees, Dr Frederick Arnold 1847–1921

Born Leeds 20 January 1847, died Leeds 17 September 1921; FLS 1872.

F.A. Lees was a Leeds Doctor and is probably the most famous of the Yorkshire botanists. His best known work is the *Flora of West Yorkshire* (Lees, 1888) but although this covers Yorkshire Bowland (now in Lancashire) he published almost nothing of this area in his book. However he did publish an earlier note on the area (Davis and Lees, 1878). A few records from northern Lancashire are mentioned in other publications and a few voucher specimens are contained in his herbarium at Cliffe Castle Art Gallery & Museum, Keighley. He also wrote a note about the flora of Bare near Morecambe (Lees, 1899) and on other West Lancashire plants (Lees, 1900). A great deal is written about Lees and Coles (2011) provides a synopsis.

Lewis, John Harbord 1848–1906

Born West Derby, Liverpool 4th Q. 1848, died Liverpool 2nd Q. 1906; FLS 1874.

John Harbord Lewis lived all his life in Liverpool where he was a boot and shoe dealer before becoming a licensee of a public house. His main collection is at the Manchester Museum but his gatherings are found in many other herbaria (Kent and Allen, 1984). He contributed a few North Lancashire records to Watson's *Topographical Botany* (Watson, 1883). See also Hancock and Pettitt (1981) and Desmond (1994).

Linton, Rev. Edward Francis 1848–1928

Born Diddington, Huntingdonshire 16 March 1848, died Southbourne, Hampshire 9 January 1928; FLS 1914.

E.F. Linton was curate at St Paul's Church, Preston for a few years in the 1870s. During this time he compiled a list of plants for V.C. 60 (Linton, 1875) but unfortunately without localities. He also wrote a note on West Lancashire plants (Linton, 1900) and contributed to Watson's *Topographical Botany* (Watson, 1883) and Wheldon and Wilson (1907). For further details see Desmond (1994).

Livermore, Leonard Albert 1919–1994

Born Lancaster 7 January 1919, died Lancaster 5 August 1994.

Len Livermore was an industrial chemist and apart from service through World War 2, lived all his life in Lancaster. On retirement in 1980 he embarked with his wife, Pat, (Plate 24) on a detailed tetrad survey of the flowering plants and ferns of North Lancashire (Livermore and Livermore, 1987a). Further reports followed covering the coast, dismantled railways, canals and urban Lancaster (Livermore and Livermore, 1987b, 1989, 1990a, 1990b and 1991a). In his later years he took a particular interest in difficult groups and amassed a large collection of ferns (especially *Dryopteris affinis s.l.*), hawkweeds and dandelions. The latter were all named by referees but only a few of the hawkweeds and ferns have been checked. These 'critical' collections came to World Museum, Liverpool but unfortunately most of the dandelions had been destroyed by insects. Fortunately the label data survived and forms the basis of the account in Greenwood (2012). Livermore did, however, publish an account of the dandelions in North Lancashire (Livermore and Livermore, 1991b) pointing out the richness of the area but the records were unlocalised. His other collections are at Tullie House Museum and Art Gallery, Carlisle. In addition, Len and Pat Livermore published a number of other notes on the North Lancashire flora. These included notes on *Allium sativum* (1992a), *Azolla filiculoides* in the Lancaster Canal (1988), an unknown *Sorbus* from Lancaster (1992b) and on white-flowered forms of some North Lancashire plants (1991c). All the records collected by Len and Pat Livermore are incorporated into the North Lancashire Flora data-base. Whilst Len took the lead in vascular plant recording, Pat was a well-known mycologist and photographer. She was a prominent member of the British Mycological Society and was their foray secretary for several years. She was the lead author for the report of the Fungi of Gait Barrows National Nature Reserve (Livermore and Livermore, 1987b). Her fungal collections are at the Royal Botanic Gardens, Kew. Pat Livermore, née Christie was born in Lancaster in 1928 and survived Len until 2010. Together they worked as a team, initially in general natural history but later as botanist and mycologist. They were prominent members of the Lancashire Wildlife Trust to which both left legacies. An obituary of Len was published in *Watsonia* (Rich, 1995).

Marshall, Rev. Edward Shearburn 1858–1919

Born London 7 March 1858, died Tidenham, Gloucestershire 25 November 1919; FLS 1887.

Rev. E.S. Marshall was a well-known British botanist and visited North Lancashire on a few occasions. He contributed to Wheldon and Wilson (1907) and wrote a note about additions to the flora (Marshall, 1896). See also Desmond (1994).

Mason, Rev. William Wright 1853–1932

Born Wainfleet St Mary, Lincolnshire 31 October 1853, died Louth, Lincolnshire 11 August 1932.

Rev. William Wright Mason came from Lincolnshire but between about 1894 and 1913 lived in Bootle, Lancashire. Here he became friends with J.A. Wheldon (Wheldon, 2011) and together they went on botanical excursions including one to Dolphinholme and Abbeystead in the Wyre valley in 1901. He is acknowledged as having contributed to the *Flora of West Lancashire* (Wheldon and Wilson, 1907). Desmond's reference (1994) to a manuscript with Liverpool Botanical Society has not been traced. For further details see Gibbons (1975) who published a portrait, A.A.D[allman] (1937) also with a portrait and Desmond (1994).

Melvill, James Cosmo 1845–1929

Born Hampstead, London 1 July 1845, died Meole Brace, Shropshire 4 November 1929; FLS 1870.

James Cosmo Melvill and Charles Bailey (see above) are credited with founding the Manchester Museum Herbarium. Both were Manchester business men but Melvill did not live in North Lancashire. However he is credited with records in Watson (1883) and Petty (1902). It appears he botanised in the Silverdale area in 1868. See Desmond (1994).

Mills, Professor John Norton 1914–1977

Born Kings Norton, Birmingham 3rd Q. 1914, died Caernarvon (in a climbing accident) December 1977 (Registered 1st Q. 1978).

John Mills was an animal physiologist and at his death held the Brackenbury Chair of Physiology at Manchester University. His hobbies were climbing and botanising and in this latter capacity contributed many hawkweed records to the *Flora of North Lancashire* (Greenwood, 2012). His collections are at the Manchester Museum and an obituary was published in *Watsonia* (Valentine, 1978).

Milne-Redhead, Richard 1828–1900

Born Manchester 16 January 1828, died Clitheroe, Lancashire 24 February 1900; FLS 1865.

Richard Milne-Redhead was a Manchester barrister, botanist and horticulturist. He bought a property at Holden Clough, Bolton-by-Bowland, giving rise to a nursery garden, which is still in business today. His local flora observations made in the 1870s are recorded as annotations in a copy of Bentham's *Handbook of the British Flora* (Bentham, 1865) now in the possession of Geoff Morries of Newton-in-Bowland. They are particularly noteworthy for recording the origin in the area of Wood Ragwort (*Senecio ovatus*) that dominates the vegetation of some parts of the area, especially in Gisburn Forest. (Greenwood, 2012). See also Desmond (1994)

Moss, John 1853–1925

Born Altrincham, Cheshire 1st Q. 1853, died St Michael's-on-Wyre, Lancashire 15 April 1925.

John Moss was born in Altrincham but came to St Michael's-on-Wyre as a teacher at the local school (Plate 29). Here, with the vicar, Ven. Phipps J. Hornby, compiled a list of plants found in the St Michael's area. (Hornby and Moss, 1925) and formed a museum of local interest. The list was based on a project organised by John Moss in about 1900 in which pupils at the school contributed records for every month of the year. In this project records were made from local farms with several interesting discoveries. The MS with recorders and locality details is now with the Lancashire County Museum Service.

Motley, James 1822–1859

Baptised Leeds 25 May 1822, died (murdered) Bangkal, Borneo 1859.

James Motley was a mining engineer and spent most of his life in south-eastern Asia. Riddelsdell (1902) published the localities for a few V.C. 60 plants found in the Motley herbarium at the Royal Institution of South Wales, Swansea. In due course the collections of the Institution passed to the Swansea Museum. See also Desmond (1994).

Newdigate, Fr Charles Alfred 1863–1942

Born Shardlow, Derbyshire 2nd Q. 1863, died Chipping Norton, Gloucestershire 2nd Q. 1942.

Fr Charles Newdigate was a Priest and teacher. According to Wheldon and Wilson (1907) and Desmond (1994) he collaborated in publishing a *Flora of Stonyhurst* (Anon., 1891) whilst he was at Stonyhurst College. However, according to Fr F.J. Turner, librarian at Stonyhurst, Newdigate's collaborator was Fr H. Wright (Turner, c. 1986). The herbarium at Stonyhurst contains a number of voucher specimens for the *Flora of Stonyhurst*.

Newton, Dr Jennifer Margaret (née Clapham), MBE 1937–2013

Born Oxford 24 February 1937, died Hornby, Lancashire 2 March 2013

Jennifer Newton (Plate 25) was born in Oxford, one of two daughters of the late Professor A.R. Clapham, one of the authors of the *Flora of the British Isles* (Clapham, Tutin and Warburg, 1952). Through several editions this became the standard work for many years for identifying British flowering plants and ferns. With this background it is not surprising that she became interested in natural history at an early age and at eleven became the grasshopper recorder for Sheffield. She went to Cambridge University where she studied botany and zoology before taking a teacher's Diploma. After a short spell teaching she obtained a PhD in plant physiology in the Agriculture Department of Oxford University.

In 1964 she married David Newton and following a period in California returned to England where David took up an appointment at Lancaster University in 1968. They were both soon involved in the cultural life of the region.

For Jennifer this meant pursuing her interests in natural history and apart from two years in Switzerland, carried out botanical surveys during the 1970s and '80s of many of the woodlands in Lancaster District. She also became involved with the Lancashire Wildlife Trust and was instrumental in acquiring several Lune valley woodlands as Wildlife Trust nature reserves. Similarly she was very much involved in the protection of Warton Crag. Warton Crag and the limestone hills of Morecambe Bay inspired her to take an interest in various invertebrate groups and she carried out long term monitoring of butterflies on Warton Crag. However she did not forget plants and undertook detailed recording for the B.S.B.I. monitoring scheme 1987–1988 (Rich and Woodruff, 1990; Palmer and Bratton, 1995) and contributed to the *New Atlas of the British Flora* (Preston, Pearman and Dines, 2002). She joined the B.S.B.I. in 1978 and as her invertebrate interests developed she became an expert in spiders. She joined the British Arachnological Society in 1992 and her obituary (Priestley, 2013) lists her publications. An obituary was also published in the *Guardian* on 27 April 2013.

Although she maintained that she was primarily an ecologist she exercised a critical approach to plant and animal identification. Apart from her projects she always kept an eye open for interesting plants that she and her wide circle of friends had recorded. Her contribution to the documentation of the Lancashire flora is acknowledged in the *Flora of North Lancashire* (Greenwood, 2012).

Apart from her natural history and nature conservation interests she was an accomplished musician. She had been first clarinet in the National Youth Orchestra and subsequently taught the clarinet as well as playing in the Lancaster Haffner Orchestra.

In 2007 she was awarded the MBE for services to nature conservation.

Oddie, Bernard 1909–1998

Born Clitheroe, Lancashire 7 February 1909, died Clitheroe 27 October 1998.

Bernard Oddie's father and family worked in local cotton mills. However Bernard (Plate 26) left school at 14 and worked through night school and college to progress from a carpenter to a teacher of woodwork and building science. He lived in West Bradford where he enjoyed walking in the countryside and took a keen interest in natural history. Bernard volunteered to record the flora of hectad SD64 and most of the records gathered in this hectad for the *Flora of North Lancashire* (Greenwood, 2012) in the late 1960s were his. Amongst his botanical papers, at World Museum, Liverpool, is a diary detailing his walks in the Bowland Fells from 1944 to 1995. These illustrate his great love of the area and its natural history.

Pearce, Edythe Mary (née Radcliffe) 1923 –2012

Born Grantham, Lincolnshire 2 April 1923, died Newark, Nottinghamshire 29 February 2012

Edythe Pearce undertook the initial surveys of tetrads SD33 J, P, N, S, T, U and SD34 K. She lived in Poulton-le-Fylde and was a keen recorder for the *Flora of North Lancashire*, often in the company of the Rev. C.E. Shaw (see below). She was also a bird recorder in the Singleton area. After marrying and living in Bolton for a while she moved to Poulton-le-Fylde, possibly after her husband's death. Sometime after 1970 she moved away from Poulton and lived in Newark where she died.

Pearsall, William Harrison 1860–1936

Born Stourbridge, Worcestershire 6 June 1860, died Matfield, Kent 12 August 1936.

W.H. Pearsall was a school teacher, mostly at Dalton-in-Furness and a well-known Cumbrian botanist. He was also secretary of the Botanical Society and Exchange club following the death of George Claridge Druce in 1932. He took a particular interest in aquatic vegetation especially in the Lake District but in North Lancashire published a single paper on Hawes Water (Pearsall, 1916). On retirement in 1925 he moved to Kent. Several obituaries were written (see Desmond, 1994) amongst which was one published in the *North Western Naturalist* (Anon., 1936).

Pearson, James 1823–1877

Born Little Eccleston, Lancashire 4 September 1823, died Rochdale, Lancashire 1st Q. 1877.

James Pearson was born at Little Eccleston in the Fylde but baptised at Lytham. Initially he was a teacher but later became a printer and book-seller in Rochdale where he died in 1877. He was a correspondent of C.J. Ashfield (see above) contributing many records, especially from the Garstang area, for his *Flora of Preston and Neighbourhood* (1858, 1860, 1862, 1864).

Peel, Mary Nina 1877–1933

Born Lancaster 3rd Q. 1877, died Canterbury, Kent 1st Q. 1933.

Mary Peel lived at Knowlmere Hall, Newton-in-Bowland, her parents' family home, until her marriage in 1915 to Nicholas Assheton. She contributed to J.F. Pickard's work on the Bowland flora (see below) and published two papers of her own (Peel, 1913a & b).

Petty, Samuel Lister *c.* 1860–1919

Born Rio de Janeiro *c.* 1860, died Ulverston, Cumbria 10 May 1919.

Samuel Lister Petty was born in Rio de Janeiro but lived and died in Ulverston. It is not known how he acquired his wealth but census records show he lived on his own means. He wrote papers on the flora of Leck and Silverdale (Petty, 1893; 1902) and contributed records to Wheldon and Wilson (1907). See also Desmond (1994).

Pickard, Joseph Fry 1876–1943

Born Silverdale, Lancashire 3 April 1876, died Leeds 18 February 1943.

Joseph Pickard was born in Silverdale but lived in Leeds. He was a Quaker, draper and tea dealer. He became interested in botany as a boy (D[allman], 1945) and when his parents moved to Newton-in-Bowland, possibly to the Heaning, in about 1890 he became more deeply inter-ested in the subject and published his first notes on the flora in 1893 aged 17 (Pickard, 1893). His papers in 1901 and 1902 (Pickard, 1901, 1902) provide early contributions to the flora of this area whilst his herbarium at Leeds University provides voucher specimens and additional records for the area. Pickard was encouraged in his hobby by his sisters and relatives who are responsible for records from Silverdale. Unfortunately it has proved difficult to identify these recorders. They include siblings Miss E. Pickard (Eliza?), Miss K. Pickard (Catherine?), E.M. Pickard (Esther Maria?), E.S. Pickard (Edward Smith Pickard) and G.S. Pickard, who may have been George Smith Pickard, possibly a cousin. All are acknowledged by Wheldon and Wilson (1907). Amongst his obituaries is one published with portrait in the *North Western Naturalist* (D[allman], 1945).

Ratcliffe, Alice Ellen 1904–1974

Born Haslingden, Lancashire 28 April 1904, died Blackpool, Lancashire 3rd Q. 1974.

Alice Ratcliffe lived at St Anne's and was an enthusiastic member of the Wild Flower Society. She was also a member of the Fylde Naturalists and the Botanical Society of the British Isles, which she joined in 1961. A copy of her Wildflower Society diary, with records from all over the country and her herbarium are at World Museum, Liverpool. She took particular responsibility for recording tetrads in hectad SD32 for the *Flora of North Lancashire* (Greenwood, 2012) and most of the records from the 1960s are hers. However she contributed records from many other areas especially around Blackpool, in the Fylde and Preston area. In the 1960s tips and other areas of waste ground provided a rich hunting ground for introductions and casual species. These Alice Ratcliffe noted and often had refereed, making a particularly valuable contribution to the understanding of the flora of North Lancashire.

Ray, Rev. John 1627–1705

Born Black Notley, Essex 29 November 1627, died Black Notley 17 January 1705; FRS 1667.

Apart from records in his publications John Ray passed through northern Lancashire on his way back from the Isle of Man in *c*. 1660 (Raven, 1942). The few records he made on this journey suggest that he travelled from

Wardleys on the north bank of the Wyre estuary (the main port at this time) to Garstang via Pilling. In the seventeenth century this was a hazardous journey across undrained bogs and fens in a landscape that it is difficult to comprehend today. Thus his disputed record of Marsh Fleawort (*Tephroseris palustris*), from ditches at Pilling, a plant which he would have known from Cambridgeshire, is probably correct (Greenwood, 2012). See also Desmond (1994).

Riddelsdell, Rev. Harry Joseph 1866–1941

Born Hackney, London 25 July 1866, died Surrey 17 October 1941

It is not thought that Rev. H.S. Riddelsdell botanised in northern Lancashire but he published a paper on the Motley herbarium containing a few Lancashire records. (Riddelsdell, 1902). See also Desmond (1994).

Robinson, Ellen Cartmell 1855–1933

Born Preston, Lancashire 1st Q. 1855, died Preston 4th Q. 1933.

Ellen Robinson was secretary of the botanical section of Preston Scientific Society at the time of the publication of their Flora (Preston Scientific Society, 1903). In the 1881 census she was an assistant in her father's bookshop. However she is not identified in subsequent censuses.

Rowse, Joseph Stephen 1842–1927

Born Briarfield, Lancashire 2nd Q. 1842, died Bury, Lancashire 2nd Q. 1927.

Wheldon and Wilson (1907) acknowledged the presence of Rowse specimens from North Lancashire in Wheldon's herbarium but none were listed in the appendix to Wheldon (2011). Joseph Rowse was a clerk in a cotton mill in Heywood in the 1901 census but is best remembered for a series of articles in the *Heywood Advertiser* in the early 1900s providing biographies of Lancashire and Cheshire artisan naturalists. Many of these people were near contemporaries and his biographies were often based on personal recollections.

Salmon, Charles Edgar 1872–1930

Born Reigate, Surrey 22 November 1872, died Reigate 1 January 1930; FLS 1902.

Charles Edgar Salmon was an architect and Surrey botanist. He provided records for Wheldon and Wilson (1907) and with H.S. Thompson (1902) published a note on North Lancashire plants. See also Desmond (1994).

Searle, Henry (Harry) 1844–1935

Born West Fen, Cambridgeshire, December 1844, died Oldham, Lancashire 26 January 1935.

Harry Searle was a prominent South Lancashire botanist but is credited with a few records from V.C. 60 by Wheldon and Wilson (1907). For an obituary, with portrait, see D[allman] (1939).

Simpson, Rev. Samuel 1802–1881

Baptised Lancaster 22 December 1802, died Chester 2 July 1881.

Samuel Simpson (Plate 12) was a Lancaster solicitor having served his articles in Liverpool. Also in Liverpool at that time was his sister Mary Maria who was training as an artist and there they met Henry Borron Fielding (see above). Fielding and Mary Simpson married and moved to Stodday near Lancaster (see above). The Fieldings and Samuel Simpson were active botanists in Lancaster during the 1830s and 1840s with Samuel Simpson being the more outgoing. He represented the Botanical Society of London in the region. Many specimens from his herbarium are at Oxford University but probably represent material he gave to Fielding. There are also Simpson specimens in H.C. Watson's herbarium at Kew but his main collection is lost. He published little but the odd record is found in the literature. Samuel Simpson was a prosperous solicitor and built a substantial house then on the outskirts of Lancaster. It is now a restaurant known as Greaves Park. However after his marriage to Ann Atkinson in 1844 he took holy orders and became Chaplain at St Thomas', Douglas, Isle of Man in 1851 where he remained until 1867. After leaving Lancaster he took no further interest in botany. Nevertheless he wrote two letters to Sir William Hooker, Director of the Royal Botanic Gardens,

Kew. The first concerned a manuscript of J.E. Stocks, a well-known Indian botanist (Desmond, 1994) and the second concerning the herbarium of James Motley (see above). In both cases his good offices had been sought by relatives of Stocks and Motley, members of his Douglas congregation, concerning the future of their collections. After leaving the Isle of Man he resided in Bristol and Chester where he died. He is buried in an unmarked grave at Lancaster Cemetery. An obituary and report of his funeral was published in the *Lancaster Gazette* for 9th July 1881.

Sledge, Dr William Arthur 1904–1991

Born Leeds 14 February 1904, died Leeds 15 December 1991.

W.A. Sledge was a Leeds botanist and had an academic career at Leeds University. He was editor of the *Naturalist* from 1943 until 1975 and Botanical Society of the British Isles recorder for V.C. 64, Mid-west Yorkshire, amongst others, from 1949 until 1987. He was elected an honorary member in 1987. As vice-county recorder for V.C. 64 he was responsible for recording in that part of Bowland now in Lancashire but formerly in Yorkshire. However apart from this long period of involve-ment few records can be attributed to him although he led a few field trips in the region. Further details of his career were published by Coles (2011) and an obituary was published by Abbott (1992).

Steeden, Charles Frederick 1922–2010

Born St Anne's, Lancashire 7 November 1922, died St Anne's 3 August 2010

Charles F. Steeden (Plate 28), known as Derek, was born and lived all his life in St Anne's. During World War 2 he was a radio operator with the R.A.F. stationed in India. On leaving the R.A.F. he became a keen radio ham ('G2HCP') contacting people all over the world for many years.

His early employment was with the Blackpool department store, RHO Hills, where he became general manager. Following a home secre-tarial study course he qualified and became a Fellow of the Institute of Chartered Secretaries (FICS). He then became company secretary of the William Birtwistle Group of companies.

In the 1960s he attended a countryside appreciation course run by local naturalist and teacher, Gordon Stead, later chair of the Lancashire Wildlife Trust. This provided the inspiration to develop a keen interest in many branches of natural history, which he shared with Jeremy, one of his three sons. There followed many years of observing, studying, photographing (he was a member of Lytham St Anne's Photographic Society) and recording wild life especially in the Fylde. He joined many societies including the Fylde Naturalists in 1969 where he initiated and produced with Jeremy the *Fylde Naturalist* (1975–1984), The Wild Flower Society (1970) becoming a branch secretary (1979–1992), British Butterfly Conservation, British Naturalists' Association, Lancashire Wildlife Trust, Preston Scientific Society (1971), Botanical Society of the British Isles (1972) and the Lancashire and Cheshire Entomological Society (1974). He was also a member of the Raven Society and an account of all their members is published by Underwood (1996).

As time passed his recording interests became more structured with a desire to understand the frequency and distribution of plants and animals particularly in the Fylde. Botanically, Alice Ratcliffe (see above) was a great help. As recording in North Lancashire developed he and Jeremy recorded the distribution of many plant species submitting an annual report over many years of the more interesting observations. This forms an invaluable data-base and the information is incorporated into the *Flora of North Lancashire* (Greenwood, 2012). Voucher specimens are deposited at World Museum, Liverpool.

Derek and Jeremy contributed to many other projects. These include *A Flora of Cumbria* (Halliday, 1997), *The Flowering Plants and Ferns of North Lancashire* (L.A. & P.D. Livermore, 1987a), BSBI Monitoring Scheme (Rich and Woodruff, 1990; Palmer & Bratton, 1990) and the *New Atlas of the British Flora* (Preston, Pearman and Dines, 2002).

Derek and Jeremy were also keen entomologists, submitting many butterfly and moth records to national schemes. Their work as recorders for West Lancashire (V.C. 60) is recognised in Heath (1989).

Following early retirement he started selling second-hand books by mail order catalogue, building up an extensive customer base of field naturalists and taking great pleasure in sourcing their desiderata for many years.

He was happily married for over 60 years and over his long and active life Derek became one of the best all-round naturalists in the region.

Thompson, Harold Stuart 1870–1940

Born Bridgwater, Somerset March 1870, died Bristol 3 March 1940; FLS 1901.

Harold Thompson was a well-known botanist but it is not thought he contributed any records to the North Lancashire flora apart from the paper he wrote with C.E. Salmon in 1902 (see above).

Thornber, Rev. William 1803–1885

Born Poulton-le-Fylde, Lancashire 2 December 1803, died Stafford 4th Q. 1885.

William Thornber was a clergyman of Poulton-le-Fylde but had many interests. In his account of *Blackpool and its neighbourhood* (Thornber, 1837) he included a list of plants. However it is believed he had little or no botanical expertise and it is not known who compiled the list. Further details about Thornber's life can be found in Clarke (1923) and Stott (1985).

Toohey, Fr Matthew 1854–1926

Born Co. Clare, Ireland 21 April 1854, died St Helens, Lancashire 26 April 1926.

Fr Matthew Toohey was a Jesuit Priest who was Chaplain at Whiston Institution, Prescot, Lancashire, 1906–1926. Whilst there he botanised in V.C. 60, especially around Blackpool and is acknowledged by Wheldon and Wilson (1925). An obituary was published in *The North Western Naturalist* (Anon., 1926).

Turner, Dawson 1775–1858

Born Great Yarmouth, Norfolk 18 October 1775, died London 20 June 1858; FRS 1802, FLS 1797.

Dawson Turner was a banker and one of the best known of the early British botanists. With L.W. Dillwyn (see above) he published *Botanists' Guide through England and Wales* (Dillwyn and Turner, 1805), which contained a few North Lancashire records. For further information see Desmond (1994).

Underhill, Alan James 1936–1999

Born Smethwick, Staffordshire 3rd Q. 1936, died Birmingham 8 April 1999.

On retirement Alan Underhill devised a project of visiting and making notes on rare species throughout the country. He visited Lancashire and left detailed locality notes on a few species. A short obituary was published by Pogson (1999).

Walters, William *fl.* 1930s

William Walters contributed several records to W.M. Gibbs' MS 'Flora of Preston 1933–39', especially from the Ashton area of the city. He lived at 35 Hull Street but otherwise nothing is known about him although by 1939 Gibbs reported that he was in ill health and not able to contribute much information.[18]

Waterfall, Charles 1851–1938

Born Leeds 7 January 1851, died Chester 26 January 1938; FLS 1911.

Charles Waterfall was a Quaker botanist who lived in Hull but moved to Chester in 1910. He recorded Parsley Fern (*Cryptogramma crispa*) on Clougha (8–10 Plants) in 1868 and on Hard Times, Lancaster Moor, Lancaster (1 plant), now a park, in 1871 (Waterfall, 1920). A.A. Dallman published an obituary (D[allman], 1938) and Waterfall's herbarium is at Sheffield University.

Watson, Hewett Cottrell 1804–1881

Born Firbeck, Yorkshire 9 May 1804, died Thames Ditton, Surrey 27 July 1881; FLS 1834.

Hewett Cottrell Watson was one of the leading nineteenth-century botanists. He is best known for his work in plant geography and for devising the vice-county system for showing the distribution of plants. Before moving to Thames Ditton he lived for a short time in the Liverpool area not far from where his father was a Cheshire lawyer. Apart from his own records he received data from many correspondents, including Samuel Simpson of Lancaster (see above). For further information see Desmond (1994).

Watts, William Marshall 1844–1919

Born Boston, Lincolnshire 2nd Q. 1844, died Rochford, Essex 13 January 1919.

William Watts was a science master at Manchester Grammar School and Giggleswick, 1868–1904. He published a 'School Flora' in 1887 that ran into many editions. The list of plants for Rossall in the 1928 edition relied on a school teacher, possibly Harold Atkinson (see above), who taught there in 1901. See Desmond (1994).

Weld, John *c.* 1813–1888

Born *c.* 1813, died Clitheroe, Lancashire 4th Q. 1888; Deputy Lieutenant for Lancashire and JP.

John Weld was a landowner and lived at Leagram Hall, Chipping. He was a man of many talents but his diaries, now at the Harris Museum and Art Gallery, Preston[17], reveal a detailed knowledge of the nineteenth-century landscape and natural history, especially birds, of his estate and neighbouring lands. The diaries contain a few botanical references. An obituary was published in the *Blackburn Standard* for 1 December 1888.

West, William 1848–1914

Born Leeds 22 February 1848, died Bradford 14 June 1914; FLS 1887.

Fisher in his account of the Lancashire flora (Fisher, 1906) acknowledges the help of William West of Bradford, a pharmaceutical chemist. However it is not clear what contribution he made to the study of the Lancashire flora. See also Coles (2011) and Desmond (1994).

Western, William Henry 1871–1948

Born Southport, Lancashire 27 June 1871, died Darwen, Lancashire 4 January 1948.

William Western was a regular contributor to the *Blackburn Times* and founded the *Lancashire* (later *Lancashire and Cheshire) Naturalist*. He joined the printing staff of a Darwen newspaper and later became a partner. He provided a few records to Wheldon and Wilson (1925). An obituary was published by Norman Ellison (1948).

Wheldon, James Alfred 1862–1924

Born Northallerton, Yorkshire 26 May 1862 died Liverpool 28 November 1924; FLS 1901.

J.A. Wheldon was a pharmacist and spent most of his working life from 1891 as pharmacist to Liverpool Prison. It is difficult to know when Wheldon became interested in the flora of Lancashire north of the R. Ribble. He met his collaborator and co-author of the *Flora of West Lancashire*, Albert Wilson (see below), in about 1884 and his V.C. 60 records and herbarium start at about the same time. He made many trips with Albert Wilson not only in North Lancashire but elsewhere and they became life-long friends. Their Flora was a model of its kind. The topographical description of V.C. 60 is detailed and could only have been written by personal exploration. Their chapters on meteorology and interpretations of plant distribution (perhaps both largely by Albert Wilson) are pioneer works when subjects such as phytogeography and ecology were still in their infancy. However they reflect their personal interests and add greatly to the Flora's value. Once the Flora was published Wheldon made only a few more visits to V.C. 60, mostly in 1910. Wheldon was particularly interested in bryophytes and the Flora and his publications reflect this. Alfred Wheldon was one of the finest amateur botanists of his day and there are several notes and biographies (see Desmond, 1994) but perhaps the most personal and readable is by his great grandson, William Peter Wheldon (Wheldon, 2011). Wheldon's herbarium is at the National Museum of Wales.

Whellan, James Arden 1915–1995

Born West Derby, Liverpool 4 April 1915, died Sydney, Australia 17 August 1995.

James Whellan knew H.E. Bunker (see above), with whom he collaborated, published two papers in *The North Western Naturalist* (Whellan 1942, 1954) and contributed to Plant Records in *Reports of the Botanical Society and Exchange Club*. He was a member of the Botanical Society of the British Isles and its predecessor from 1942 until 1984. However Whellan was primarily a tropical entomologist and spent most of his working life on pest control work in former UK colonies. In this capacity he had a distinguished career and published several papers and reports (Jerry

Cooper and Margaret J. Haggis, pers. comm.). Whilst in Africa he became interested in succulent plants, especially the genera *Aloe*, *Euphorbia* and *Brachystelma*. On retirement he moved to Sydney, Australia where he died in 1995. An obituary was published in the *British Cactus and Succulent Journal* (PED, 1996).

Whiteside, Robert 1866–1960

Born Lancaster 3rd Q. 1866, died Lancaster 9 November 1960.

Robert Whiteside lived all his life in Lancaster where he had a laundry business. He was President of the Lancaster Horticultural Society for many years but was primarily a pteridologist. He founded the North British Pteridological Society in 1891. His collections are at the Natural History Museum, London and a few records in Greenwood (2012) are attributed to him. See Desmond (1994).

Whittle, Peter Armstrong 1789–1867

Born Inglewhite, Lancashire 9 July 1789, died Liverpool 7 January 1867.

Peter Whittle was born at Inglewhite near Preston and became a printer and publisher in Preston. He published histories of Lytham and Blackpool amongst others with lists of plants. It is not clear to what extent he compiled these himself or relied on others. Several botanists meeting in public houses are listed for Preston but nothing of their work has been traced. He intended publishing a *Flora Prestoniensis* but this was never produced and a manuscript has not been traced. See also Desmond (1994).

Wilson, Albert 1862–1949

Born Garstang, Lancashire 12 October 1862, died Priest Hutton, near Carnforth, Lancashire 15 May 1949; FLS 1900.

Albert Wilson, like his collaborator and co-author of the *Flora of West Lancashire*, J.A. Wheldon, became interested in botany at an early age. He was the second son of Charles and Susannah Wilson. Charles was a Quaker and tailor who, after living in Preston for some years, moved to Calder Mount, Barnacre, Garstang. Here Susannah Wilson, in particular, encouraged her son Albert, in his botanical interests. She is probably the

Mrs C. Wilson who found Mousetail (*Myosurus minimus*) at Slyne-with-Hest *c.* 1860. Albert Wilson was a pharmacist and became a partner in a firm in Bradford where he had been apprenticed. He married Alice Mary Thorpe in 1890 and she together with his brother Sydney and later his son Howard all helped and encouraged Albert in his botanical interests. Although encouraged as a boy by his parents when they lived at Garstang his major work on the V.C. 60 flora was accomplished when he was living in Ilkley. It was not until he retired from business at the age of 50 in 1912 that he returned to live in Garstang. After his mother died in 1916 he moved to Bentham and then Sedbergh. Later he moved to North Wales before moving to live at Priest Hutton following his wife's death in 1945. Thus, whilst his parental home was in Garstang in the years devoted to the Flora project he lived elsewhere. No doubt accommodation was always available but frequently with his friend J.A. Wheldon the years between 1898 and 1907 were spent exploring the often remote parts of the vice-county. Wilson in particular was a keen walker and a journey of 20 to 30 miles on foot was no problem. The resulting *Flora* was a masterpiece in coverage and for its introductory chapters (see Wheldon above). Wilson was also a photographer and the fifteen topographical plates are an invaluable record. In addition to numerous papers Wilson also wrote the *Flora of Westmorland* (Wilson, 1938). His herbarium is at the Yorkshire Museum but he also wrote notes on other places he visited. Many of these are for localities in North Wales. Many notes are contained in exercise books in the Dallman archives at World Museum, Liverpool. Also at Liverpool is a manuscript 'Flora of Priest Hutton'. Albert Wilson wrote an account of his life and botanical exploits with a portrait, which was published after his death (Wilson, 1953). An obituary was published in 1950 with a list of his publications (Lousley, 1950).

References, Bibliography and Notes

References and Bibliography

Abbott, P.P. (1992). William Arthur Sledge (1904–1991) in Obituaries. *Watsonia*, **19**: 166–168.

Abbott, P.P. (2005). *Plant Atlas of Mid-west Yorkshire.* Yorkshire Naturalists' Union. Kendal.

Allen, D.E. (1976). *The Naturalist in Britain. A social history.* Allen Lane. London.

Allen, D.E. (1986). *The Botanists.* St Paul's Bibliographies. Winchester.

Allen, D.E. (2001). *Naturalists and Society. The culture of natural history in Britain, 1700–1900.* Various series. Ashgate Publishing. Aldershot.

Allen, D.E. (2007). Vera Gordon 1918–2006 in Obituaries. *Watsonia*, **26**: 511–526.

Anon. (1792). *Universal British Directory – Warrington.* London.

Anon. (*c.* 1829). *History of Lytham in the Hundred of Amounderness in the County Palatine of Lancashire.*

Anon. (1830). *A new description of Blackpool in the parish of Bispham, Hundred of Amounderness, County Palatine of Lancashire.* Marina Press. Preston.

Anon. (1886). Flora of Stonyhurst District. *Stonyhurst Magazine*, May 1886, pp 3–12.

Anon. (1891). *Flora of the Stonyhurst District.* 2nd ed. Parkinson and Blacow. Clitheroe.

Anon. (1926). Matthew Toohey (1854–1926) in Obituary. *The North Western Naturalist*, **1**: 153–154.

Anon. (1936). William Harrison Pearsall (1860–1936) in Obituaries. *The North Western Naturalist*, **11**: 375–376.

Ashfield, C.J. (1858). On the Flora of Preston and its neighbourhood. Part 1. *Historic Society of Lancashire and Cheshire Transactions*, **10**: 143–164.

Ashfield, C.J. (1860). On the Flora of Preston and its neighbourhood. Part 2. *Historic Society of Lancashire and Cheshire Transactions*, **12**: 127–134.

Ashfield, C.J. (1861). List of plants found within 5 miles of Lytham on the banks of the Ribble. *Phytologist*, **5 NS**: 380–381.

Ashfield, C.J. (1862). On the Flora of Preston and its neighbourhood. Part 3. *Transactions of the Historic Society of Lancashire and Cheshire*, **14**: 1–6.

Ashfield, C.J. (1864). A list of Silverdale plants. *The Botanists' Chronicle*, **10**: 73–75.

Ashfield, C.J. (1865). On the Flora of Preston and its neighbourhood. Part 4. *Transactions of the Historic Society of Lancashire and Cheshire*, **17**: 181–186.

Atkinson, H.W. (1901). *Rossall fauna and flora* with annotations by Anon. Copy with late Mrs N.F. McMillan.

Bailey, C. (1902). ii. On the adventitious vegetation of the sandhills of St Anne's-on-the-Sea, North Lancashire (Vice-County 60). *Manchester Memoirs*, **47**: 1–8.

Bailey, C. (1907a). xi. Further notes on the adventitious vegetation of the sandhills of St Anne's-on-the-Sea, North Lancashire (Vice-county 60). *Manchester Memoirs*, **51**: 1–16.

Bailey, C. (1907b). *De Lamarck's Evening Primrose* (Oenothera Lamarkiana*) on the sandhills of St Anne's-on-the-Sea, North Lancashire.* Address at the annual meeting of the Manchester Field Club, Tuesday Evening 29th January 1907. Hinchcliffe and Co Limited. Manchester.

Bailey, C. (1910). xv. A third list of the adventitious vegetation of the sandhills of St. Anne's-on-the-Sea, North Lancashire, Vice-county 60. *Manchester Memoirs*, **54**: 1–11.

Belchem, J., ed. (2006). *Liverpool 800. Culture, character & history.* Liverpool University Press. Liverpool.

Bentham, G. (1865). *Handbook of the British Flora.* Lovell Reeve & Co. London.

Blackburn Field Club (1925). A visit to the Preston and Kendal Canal. *The Lancashire and Cheshire Naturalist*, **17**: 267–268.

Bosanquet, S.D.S. (2003). 'Monmouthshire Register of rare Bryophytes'. Unpublished Report. Dingestow Court. Monmouth.

Bowran, J.G. (1932). Rev. Alfred John Campbell FLS. *Memoirs Minutes of Primitive Methodist Conference 1931.*

Braithwaite, M. and Walker, K. (2012). *50 years of mapping the British and Irish flora 1962–2012.* Botanical Society of the British Isles. London.

Bryant, D., ed. (2008). *In the footsteps of Eric Hardy.* Hobby Publications. Southport.

Buckley, N. (1842). List of plants in the vicinity of Lytham, Lancashire. *Phytologist,* **1**: 165–166.

Burns, J. (1955). Some of the rarer plants in the Lancaster District. *New Biologian,* **4**: 9–11.

Carbis, N. (1978). *Nellie Carbis looks back.* Titus Wilson & Son Ltd. Kendal.

Clapham, A.R., Tutin, T.G. and Warburg, E.F. (1952). *Flora of the British Isles.* Cambridge at the University Press.

Clarke, A. (1923). *The story of Blackpool.* Palatine Books. Blackpool.

Clay, W.L. (1861). *The prison chaplain. A memoir of Rev. John Clay BD.* Macmillan and Co. London

Coles, G.L.D. (2011). *The story of South Yorkshire Botany and the* Flora Sheffieldiensis *of Jonathan Salt.* Yorkshire Naturalists' Union. Kendal.

Corner, B.C. and Booth, C.C. (1971). *Chain of friendship. Selected letters of Dr. John Fothergill of London 1735–1780.* The Beknap Press of Harvard University Press, Cambridge, Massachusetts and University of Oxford Press, Oxford.

Crosby, A. (1998). *A history of Lancashire.* Phillimore & Co. Ltd. Chichester.

Crosby, A. (2006). Moving through the landscape in Winchester, A.J.L., ed., *England's landscape. The North West.* Collins. London.

Crosby, A. (2012). *The history of Preston Guild.* Carnegie Publishing Ltd. Lancaster.

Crosfield, G. (1843). *Memoirs of the life of Samuel Fothergill.* D. Marples. Liverpool.

Crosfield, J.F. (1980). *The Crosfield family. A history of the descendants of Thomas Crosfield of Kirkby Lonsdale who died in 1614.* Privately published.

Cross, W. (1889). Among the Fylde flowers. *Wesley Naturalist,* **2**: 322–324.

D[allman]., A.A. (1937). William Wright Mason (1853–1937) with portrait, plate 19. *The North Western Naturalist,* **12**: 316–327.

D[allman]., A.A. (1938). Obituary. Charles Waterfall (1851–1938). *The North Western Naturalist*, **13**: 104–105.

D[allman]., A.A. (1939). A Lancashire botanist: Henry (Harry) Searle (1844–1935) and his circle with portrait. *The North Western Naturalist*, **14**: 262–269.

D[allman], A.A. (1945). Obituary. Joseph Fry Pickard (1876–1943). *The North Western Naturalist*, **20**: 285–287.

Dale, P. (*c.* 1986). A revised Stonyhurst Flora. *The Stonyhurst Magazine*, pp. 361–370.

Dalziel, N. (1993). Trade and transition 1690–1815 in White, A., ed., *A history of Lancaster 1139–1993*. Ryburn Publishing. Keele University Press.

David, R. and Winstanley, M with Bainbridge, M. (2013). *The West Indies and the Arctic in the age of sail: the voyages of* Abram *(1806–62)*. Centre for North-West Regional Studies. Lancaster University.

Davis, P. (1996). The Backhouses and their scientific pursuits. *Occasional papers National Botanic Gardens, Glasnevin*, **8**: 37–54.

Davis, J.W. and Lees, F.A. (1878). *West Yorkshire*. L. Reeve & Co. London.

Desmond, R. (1994). *Dictionary of British and Irish Botanists and Horticulturists*. Taylor & Francis and the Natural History Museum. London.

Dillwyn, L.W. and Turner, D. (1805). *The botanist's guide through England and Wales*. Phillips and Fardon. London.

Dobson, W. (1883). *Rambles by the Ribble*. 3rd Series. Preston.

Druce, C.G. (1926). *The Flora of Buckinghamshire*. T. Buncle & Co. Arbroath.

Dunlop, G.A. (1921). The occurrence of Parsley Fern (*Cryptogramma crispa* Bl.) in Lancashire and Cheshire. *Lancashire and Cheshire Naturalist*, **13**: 143.

Dunlop, M. (1936). The Fylde: phytogeography in Grime, A., ed., *A scientific survey of Blackpool and District*. British Association for the Advancement of Science. London.

Edees, E.S. (1972). *Flora of Staffordshire*. David & Charles. Newton Abbot.

Elder, M. (1992). *The slave trade and the economic development of 18th century Lancaster*. Ryburn Publishing. Halifax.

Ellison, N.F. (1948). Obituary. William Henry Western (1871–1948). *The North Western Naturalist*, **32**: 169–170.

Fisher, H. (1906). Part 2 Botany in Fowler, W. & Brownbill, J., eds, *A history of the County of Lancaster.* The Victoria history of the counties of England. Constable and Company Limited. London.

Fothergill, R. (1998*). The Fothergills. A first history.* Richard Fothergill. Newcastle-upon-Tyne.

Foley, M. (2008). Margaret Baecker: still going strong at 105. *BSBI News,* **107**: 26–28.

Fowler, W. (1881). Notes on the flora of Hodder-Dale in Short Notes and Queries. *The Naturalist,* **7**: 15–16.

France, R.S. (1931a). Notes on the flora of West Lancashire. *The North Western Naturalist,* **6**: 99–100.

France, R.S. (1931b). The changing flora of ... *Countryside,* **9**: 21. (Not seen)

France, R.S. (No date). Annotations in a copy of Wheldon and Wilson (1907). Copy at Lancaster University.

Frankland, J.N. (2001). *A Flora of Craven.* North Craven Heritage Trust. Settle.

Garlick, G.W. (1957). A saltmarsh flora of the Heysham Peninsula. *New Biologian,* **6**: 8–10.

Gerard, J. (1636). *Historie of Plants.* 2nd ed, edited by Thomas Johnson.

Gibbons, E.J. (1975). *The Flora of Lincolnshire.* Lincolnshire Naturalists' Union. Lincoln.

Gibbs, W. (*c.* 1939). MS Flora of Preston 1933–1939. Copies at Lancashire Record Office and World Museum, Liverpool.

Gordon, A. (1891). Hutton, William 1723–1815 in Lee, S. ed., *Dictionary of National Biography,* vol. **28**. Smith Elder & Co. London.

Graham, G.G. (1988). *The Flora and Vegetation of County Durham.* The Durham Flora Committee and the Durham County Conservation Trust.

Greenwood, B.D. (1977). The papers of Arthur Augustine Dallman (1883–1963). *Journal of the Society for the Bibliography of Natural History,* **8**: 176–179.

Greenwood, E.F. (2003). Understanding change: a Lancashire perspective. *Watsonia,* **24**: 337–350.

Greenwood, E.F. (2005). The changing flora of the Lancaster Canal in West Lancaster (V.C. 60). *Watsonia,* **25**: 231–325.

Greenwood, E.F. (2012). *Flora of North Lancashire.* Palatine Books for the Lancashire Wildlife Trust. Lancaster.

Groom, Q.J., O'Reilly, C and Humphrey, T. (2014). Herbarium specimens reveal the exchange network of British and Irish botanists, 1856–1932. *New Journal of Botany*, **4**: 95–103.

Hadfield, C. and Biddle, G. (1970). *The canals of North West England.* 2 vol. David & Charles. Newton Abbot.

Haley, R.A. (1995). *Lytham St Anne's. A pictorial history.* Phillimore. Chichester.

Halliday, G. (1997). *A Flora of Cumbria.* Centre for North-West Regional Studies, University of Lancaster. Lancaster.

Hancock, E.G. (1980). Biographical notes on Thomas Greenlees (1865–1949) in Short Notes. *Watsonia*, **13**: 54–55.

Hancock, E.G. and Pettitt, C.W. (1981). *Register of Natural Science Collections in north west England.* Manchester Museum for North West Collections Research Unit. Manchester.

Heath, J. (1989). *Moths and Butterflies of Great Britain and Ireland.* Volume 7 part 1. Apollo Books.

Heathcote, W.H. (1923). Additions to the Flora of Preston. *Lancashire and Cheshire Naturalist*, **15**: 140.

Henderson, Cl. W. (1962). Northern limestones. *Bulletin of the Alpine Garden Society*, **30**: 306–318.

Hill, M.O., Preston, C.D. and Smith, A.J.E. (1991–94). *Atlas of the bryophytes of Britain and Ireland.* 3 vols. Harley Books. Colchester.

Hodkinson, I.D. and Steward, A. (2012). *The three-Legged Society.* Centre for North-West Regional Studies. University of Lancaster. Lancaster.

Holt, G.O. (1978). *A regional history of the railways of Great Britain. Vol. X the North West.* David & Charles. Newton Abbott.

Holt, J. (1795). *General view of the agriculture of the County of Lancashire.* Reprinted 1969 by David & Charles Reprints. Newton Abbott.

Hornby, P.J. and Moss, J. (1925). *List of flowers, grasses & etc. gathered in a radius of two miles from St Michael's Church.* St Michael's Museum. St Michael's.

Howson, G. (2002). 5. The Lancaster Doctors: three case studies in Wilson, S., ed., *Aspects of Lancaster*: 53–62. Wharncliffe Books. Barnsley.

Hunt, D. (1992). *A history of Preston.* Carnegie Publishing in conjunction with Preston Borough Council. Preston.

Hutton, W. (1789). *A description of Blackpool in Lancashire frequented for sea-bathing.* Pearson.

Inchbald, P. (1865). The coast round Blackpool in March. *Naturalist*, **1**: 361–363.

Jackson, J.W. (1926). Obituary. William Henry Heathcote (1861–1926). *The North Western Naturalist*, **1**: 85–87.

Jenkinson, J. (1775). *A generic and specific descriptions of British Plants translated from the Genera et Species Plantarum of the celebrated Linnaeus ...together with notes and observations by James Jenkinson.* London, Kendal and Lancaster.

Kent, D.H. and Allen, D.E. (1984). *British and Irish Herbaria.* Botanical Society of the British Isles. London.

Kirkby, W. (1902). Silverdale plants in Notes on Flowering Plants. *The Naturalist*, **1902**: 316.

King, F.C. (1888). Geology, botany and physical history of the district in Smith, T.C., *A history of Longridge and District.* C.W. Whitehead. Preston. (King also prepared the drawings for the engravings.)

Lankester, E., ed. (1848). *The correspondence of John Ray.* Ray Society. London.

Lees, F.A. (1888). *The Flora of West Yorkshire.* Lovell Reeve & Co. London.

Lees, F.A. (1899). The Florula of Bare, West Lancashire. *The Naturalist*, **1899**: 299–303.

Lees, F.A. (1900). West Lancashire Indigenes. *The Naturalist*, **1900**: 3–4.

Linton, E.F. (1875). v. Tabular catalogues of common plants for Breconshire, Radnorshire, Selkirkshire and West Lancaster in Report of the Recorder for 1874. *The Botanical Locality Record Club*, **1875**: 80–86.

Linton, E.F. (1900). West Lancashire additions in Short Notes. *The Journal of Botany*, **38**: 86–87.

Livermore, L.A. and Livermore, P.D. (1987a). *The flowering plants and ferns of north Lancashire.* L.A. & P.D. Livermore. Preston.

Livermore, L.A. and Livermore, P.D. (1987b). *Fungi of Gait Barrows National Nature Reserve : A report by the North West (England) Region of the Nature Conservancy Council*, ed. N.A. Robinson. Nature Conservancy Council (Great Britain). North West Regional Office, Blackwell.

Livermore, L.A. and Livermore, P.D. (1988). *Azolla filiculoides* in the Lancaster Canal. *Pteridologist*, **1**: 214–215.

Livermore, L.A. and Livermore, P.D. (1989). *The flowering plants, ferns*

& rusts of the Lancaster Canal in the Lancaster District. L.A. & P.D. Livermore. Lancaster.

Livermore, L.A. and Livermore, P.D. (1990a). *Coastal plants and rust fungi of the North Lancashire coast.* L.A. & P.D. Livermore. Lancaster.

Livermore, L.A. and Livermore, P.D. (1990b). *Plants and rust fungi of the dismantled railway lines in the Lancaster District.* L.A. & P.D. Livermore. Lancaster.

Livermore, L.A. and Livermore, P.D. (1991a). *Lancaster's plant life – a botanical survey.* L.A. & P.D. Livermore. Lancaster.

Livermore, L.A. and Livermore, P.D. (1991b). *Taraxacum* flora of North Lancashire. *BSBI News*, **57**: 9–10.

Livermore, L.A. and Livermore, P.D. (1991c). White flowered forms of some N. Lancashire plants. *BSBI News*, **58**: 12.

Livermore, L.A. and Livermore, P.D. (1992a). *Allium sativum* L. (Garlic). *BSBI News*, **60**: 11.

Livermore, L.A. and Livermore, P.D. (1992b). An unknown *Sorbus* in Lancaster. *BSBI News*, **60**: 17.

Lousley, J.E. (1950). Albert Wilson (1862–1949) in Obituaries. *Watsonia*, **1**: 327–331. (Includes a list of publications.)

Margetts, L.J. and David, R.W. (1981). *A Review of the Cornish Flora 1980.* Institute of Cornish Studies. Redruth.

Marshall, E.S. (1896). Additions to the flora of Lancashire. *The Journal of Botany*, **34**: 136.

Marshall, J.D. (1967). *The autobiography of William Stout of Lancaster 1665–1752.* Manchester University for the Chetham Society. Manchester.

Milne-Redhead, R. (1870s). Annotations in Bentham, G. (1865). *Handbook of the British Flora.* Lovell Reeve & Co. London. In the possession of G. Morries, Newton-in-Bowland in 2008.

Mourholme Local History Society (1998). *How it was. A North Lancashire Parish in the seventeenth Century.* Carnforth.

Nottingham, L. (1992). *Rathbone Brothers. From Merchant to Banker 1742–1992.* Rathbone Brothers Plc. London.

Oakes, C. (1953). *The birds of Lancashire.* Oliver & Boyd. Edinburgh & London.

Palmer, M.A. and Bratton, J.H., eds (1990). *A sample Survey of the flora of Britain and Ireland.* UK Nature Conservation No. 8. Joint Nature Conservation Committee. Peterborough.

PED (1996). Obituary. James A. Whellan – 1915–1995. *British Cactus & Succulent Society Journal,* **14**: Newsletter, p. ii.

Pearsall, W.H. (1916). Notes on the aquatic vegetation of Hawes Water, Silverdale. *Botanical Society and Exchange Club Report,* **4**: 294.

Pearson, J. (1855). Notes on the botany of North Lancashire. *The Naturalist,* **5**: 14–16.

Pearson, J. (1874). Rare plants. *Science Gossip,* **10**: 259.

Peel, A. (1913). *The Manor of Knowlmere.* Privately published.

Peel, M.N. (1913a). Aliens and introduced plants of the upper Hodder. *The Naturalist,* **1913**: 141–143.

Peel, M.N. (1913b). The orchids of the upper Hodder valley. *The Naturalist,* **1913**: 29–32.

Percy, J. (1991). Scientists in humble life: the artisan naturalists of south Lancashire. *Manchester Region History Review,* **5**: 3–10. Manchester Polytechnic. Manchester.

Perring, F.H. and Walters, S.M., eds (1962). *Atlas of the British Flora.* Thomas Nelson and Sons Ltd for the Botanical Society of the British Isles. London.

Petty, S.L. (1893). The plants of Leck and neighbourhood, Lancashire. *The Naturalist,* **1893**: 91–102.

Petty, S.L. (1902). Some plants of Silverdale, West Lancashire. *The Naturalist,* **1902**: 33–54.

Pickard, J.F. (1893). Botany. A few notes on the flora of Newton-in-Bowland and neighbourhood. *Natural History Journal,* **17**: 31.

Pickard, J.F. (1901). Some rarer plants of Bowland. *The Naturalist,* **1901**: 37–41.

Pickard, J.F. (1902). Additions to the Bowland flora. *The Naturalist,* **1902**: 289–291.

Pogson, C. (1999). Alan Underhill in Obituary Notes. *BSBI News,* **82**: 7–8.

Preston, C.D. (2013). Following the BSBI's lead: the influence of the *Atlas of the British flora, 1962–2012. New Journal of Botany,* **3**: 2–14.

Preston, C.D., Pearman, D.A. and Dines, T.D., eds (2002). *New Atlas of the British & Irish Flora.* Oxford University Press, Oxford.

Preston Scientific Society (1903). *Flora of Preston & neighbourhood.* Preston Scientific Society. Preston.

Priestley, S. (2014). Dr Jennifer Newton MBE (1937–2013). *Newsletter British Arachnological Society,* **127**: 3–4.

'Q' (1908). The flora of St Anne's. *Lancashire Naturalist,* **1**: 161–162.

'Q' (1908). The alien flora of St Anne's. *Lancashire Naturalist*, **1**: 184.

Raistrick, A. (1968). *Quakers in science and industry*. David and Charles (Holdings) Limited. Newton Abbot.

Raven, C.E. (1942). *John Ray naturalist his life and works*. Cambridge University Press. Cambridge.

Raven, C.E. (1948). Thomas Lawson's note-book. *Proceedings of the Linnean Society of London*, **160**: 3–5.

Ray, J. (1670). *Catalogus Plantarum Angliae*. London.

Rich, T.C.G. (1995). Leonard Albert Livermore (1919–1994) in Obituaries. *Watsonia*, **20**: 449–450.

Rich, T.C.G., and Baecker, M. (1986). The distribution of *Sorbus lancastriensis* E.F. Warburg in Short Notes. *Watsonia*, **16**: 83–85.

Rich, T.C.G., Houston, L., Robertson, A. and Proctor, M.C.F. (2010). *Whitebeams, Rowans and Service Trees of Britain and Ireland*. B.S.B.I. Handbook No. 14. Botanical Society of the British Isles in association with National Museum of Wales. London.

Rich, T.C.G. and Woodruff, E.R. (1990). *The BSBI monitoring scheme 1987–1988*. 2 vols. Report for Nature Conservancy Council.

Riddelsdell, H.J. (1902). North of England Plants in the Motley Herbarium at Swansea. *The Naturalist*, **1902**: 343–351.

Rodwell, J.S., ed. (1992). *British Plant Communities*. Volume 3. Cambridge University Press. Cambridge.

Russell, C.A. (1986). *Lancashire chemist: the early years of Sir Edward Frankland*. Open University Press. Milton Keynes and Philadelphia.

Salmon, C.E. (1912). Early Lancashire and Cheshire records. *Journal of Botany*, **50**: 369–371.

Salmon, C.E. and Thompson, H.S. (1902). West Lancashire notes. *The Journal of Botany*, **40**: 293–295.

Savidge, J.P., Heywood, V.H. and Gordon, V. (1963). *Travis's Flora of South Lancashire*. Liverpool Botanical Society. Liverpool.

Secord, A. (1994a). Corresponding interests: artisans and gentlemen in nineteenth-century natural history. *British Journal for the History of Science*, **27**: 383–408.

Secord, A. (1994b). Science in the pub: artisan botanists in early nineteenth-century Lancashire. *History of Science*, **32**: 269–315.

Shakeshaft, P. (2000). *St Anne's on the Sea: a history*. Carnegie Publishing Ltd. Lancaster

Simpson, F.W. (1982). *Simpson's Flora of Suffolk*. Suffolk Naturalists'
    Society. Ipswich.
Smith, T.C. (1888). *A history of Longridge and District*. C.W. Whitehead.
    Preston.
S[tansfield]., F.W. (1931). Report of meeting to Trough of Bowland.
    British Fern Gazette, **6**: 116. (Not seen)
Stott, A., ed. (1985). *The history of Blackpool and its neighbourhood by
    William Thornber 1837*. Republished by the Blackpool and Fylde
    Historical Society. Nelson.
Sutton, C.W. (1900). Whittle, Peter Armstrong (1789–1866) in Lee, S.,
    ed., *Dictionary of National Biography*. Volume 61. Smith Elder & Co.
    London.
Thornber, W. (1837). *An historical and descriptive account of Blackpool and
    its neighbourhood*. Blackpool.
Turnbull, J.G., ed. (1951). *A history of the calico printing industry of Great
    Britain*. John Sherratt and Son. Altrincham.
Turner, F.J. (*c.* 1986). 'A preliminary Flora of the Stonyhurst District',
    1886 and 1891. *Stonyhurst Magazine*.
Underwood, R. (1996). *The Raven Entomological and Natural History
    Society – Fifty years 1946–1996*. Raven Entomological Society.
Valentine, D.H. (1978). John Norton Mills (1914–1977) in Obituaries.
    Watsonia, **12**: 187.
Waterfall, C. (1920). Occurrence of *Cryptogramma crispa* Bl. (the Parsley
    Fern) in West Lancashire in 1868 and 1871. *Lancashire and Cheshire
    Naturalist*, **13**: 46.
Watson, H.C. (1883). *Topographical Botany* 2nd ed. Bernard Quaritch.
    London.
Watts, W.M. (1928). *A School Flora*. Longmans Green and Co. London.
    Plants for Rossall information supplied by a teacher.
Wheldon, J.A. (1900). West Lancashire Flora notes. *The Naturalist*, **1900**:
    1–2.
Wheldon, W.P. (2011). *An eminent Liverpool botanist. A life of James Alfred
    Wheldon MSc, ALS, ISM (1862–1924)*. B. Wheldon. York.
Wheldon, J.A. and Wilson, A. (1900). Additions to the Flora of West
    Lancashire. *The Journal of Botany*, **38**: 40–47.
Wheldon, J.A. and Wilson, A. (1901). Additions to the Flora of West
    Lancashire. *The Journal of Botany*, **39**: 22–26.

Wheldon, J.A. (1901). Notes on the Flora of Over Wyresdale. *The Naturalist*, **1901**: 357–362.

Wheldon, J.A. (1902). West Lancashire plants. *The Journal of Botany*, **40**: 346–350.

Wheldon, J.A. and Wilson, A. (1905). Additions to the West Lancashire flora. *Journal of Botany*, **43**: 94–96.

Wheldon, J.A. and Wilson, A. (1906). Additions to the Flora of West Lancashire. *Journal of Botany*, **44**: 99–102.

Wheldon, J.A. and Wilson, A. (1907). *The Flora of West Lancashire.* Henry Young & Sons. Liverpool.

Wheldon, J.A. and Wilson, A. (1913). West Lancashire extinctions. *Journal of Botany*, **51**: 336.

Wheldon, J.A. and Wilson, A. (1925). West Lancashire Flora: notes, additions and extinctions. *Lancashire and Cheshire Naturalist*, **17**: 117–125.

Whellan, J.A. (1942). Notes on the flora of West Lancashire, Vice-county 60. *The North Western Naturalist*, **17**: 354–357.

Whellan, J.A. (1954). The present day flora of the sand dunes at St Annes, W. Lancs. V.C. 60. *The North Western Naturalist*, **2 NS**: 139–141.

Whiteside, R. (1930–50). Annotations in a copy of Wheldon and Wilson (1907). In the possession of Mrs Jones-Parry, Silverdale in the 1960s.

Whittle, P. (1821). *The history of the Borough of Preston*. Volume 1. Preston.

Whittle, P. (1837). *Topographical, statistical and historical account of Preston*. Volume 2. Preston.

Whittaker, E.J. (1986). *Thomas Lawson 1630–1691. North Country botanist, Quaker and schoolmaster*. Sessions Book Trust. York.

Wilson, A. (1938). *The Flora of Westmorland*. T. Buncle & Co. Ltd, Arbroath. Includes a portrait of the author.

Wilson, A. (1942). Druce's Comital Flora. Corrections and suggestions. *Botanical Society and Exchange Club Report*, **12**: 319–330.

Wilson, A. (1953). The Albert Wilson herbarium. *The North Western Naturalist*, **1NS**: 391–399.

Wilson, J. (1744). *Synopsis of British Plants in Mr Ray's Method*. John Gooding. Newcastle upon Tyne.

Wyatt, W. (1887). Botany of North-east Lancashire. *The Wesley Naturalist*, **1**: 179–180.

Notes

1. Tabitha Driver provided much information from the archives of Friends' House, London. In particular she searched the typescript 'Dictionary of Quaker Biography' on my behalf and checked registers for births, marriages and deaths.
2. There is a small sheet of Male-fern (*Dryopteris filix-mas*) in volume 1 of Gibson MSS (MS vol. 334) at Friends' House, London. The fern was collected by James Jenkinson in Yealand in 1744.
3. Letter from Tabitha Driver 8 June 2000 abstracting information from *Journal of the Friends' Historical Society* vol. 15 (1918).
4. 'Green Garth and the Jenkinsons: a house and its history'. A typescript by C. Shaw (no date) paginated 27–32 at Friends' Meeting House, Yealand Conyers. This refers to an eighteenth century minute book of Yealand Conyers Preparatory Meeting at the Lancashire County Record Office. This is incorporated within reference LRO. FRL. Acc. 8370.
5. Enclosure award, Yealand Conyers and Yealand Redmayne 1728. LRO. AE 5/16.
6. Leaving certificate for James Jenkinson and other archives. Archives of Lancaster Monthly Meeting of the Society of Friends. LRO. FRL. Acc, 8370. 1881.
7. Leaving certificate for George Crosfield Junr. Kendal 30th of 4th month 1777. Society of Friends archives. Manchester Central Library archives department.
8. Minutes of Preparatory Meeting Society of Friends, Warrington (Penketh) 21 of 6th month 1772. Central Library, Warrington.
9. Tithe map and schedule, Ashton-with-Stodday 1842. LRO. ARB.1/9.
10. 'Memoir of the late Henry Borron Fielding Esq FLS and GS of Lancaster 1852'. (MS Sherard 397) on loan from the Bodleian Library to the Department of Plant Sciences, University of Oxford (Plate 8). It was believed that the author of the 'Memoir' was Mrs Fielding but the handwriting does not match the signature on her will. Also the initials 'JC' written in pencil at the end suggest this was the author. This could be John Clay, who was H.B. Fielding's brother-in-law (Fielding's sister, Henrietta, married Rev. John Clay, Chaplain to Preston Gaol; Clay, 1861). The Fieldings and Clays are buried in similar graves next to each other at Churchtown near Garstang (Plate 19).

11. 'The chemistry of calico printing. History of the print works in the Manchester District from 1790 to 1840' MS by J. Graham at Manchester Central Library.

12. Box file relating to Lancaster Literary, Scientific and Natural History Society at Lancaster City Reference Library.

13. There is a reference to a talk given by John Just and a catalogue of the plants of the Kirkby Lonsdale neighbourhood in *Fourth and fifth annual reports and proceedings of the Botanical Society. Sessions 1839–40 and 1840–41.* Edinburgh.

14. Obituary of Rev. S. Simpson M.A. *Lancaster Gazette*, 9 July 1881.

15. MS 'English Flora' collected and drawn by Mary Maria Fielding. 6 volumes. Bodleian Library, Department of Special Collections and Western Manuscripts on loan to Department of Plant Sciences, University of Oxford (MS. Eng. D. 3357).

16. The Myerscough Farms Plant Survey. 2 volumes. A typescript with no author but prepared by Simon Hilton and Christopher Bruxner. Letter to E.F. Greenwood (1987) with a copy of the typescript.

17. Collection of manuscripts, diaries etc. by John Weld of Leagram Hall, Chipping 1862–1911 at Harris Museum and Art Gallery, Market Square, Preston. Further notes and diaries at Lancashire County Record Office, Bow Lane, Preston.

18. W.M. Gibbs compiled a MS 'Flora of Preston' in the 1930s, copies of which are in World Museum, Liverpool and with Lancashire County Museum Service. One of the contributors to her 'Flora' was William Walters but nothing is known about him.

Plate 1. A portion of John Speed's map of Lancashire 1610 showing the extent of mossland north of the R. Wyre. John Ray probably landed at Wardleys on the north bank of the Wyre (nearly opposite Skippool) in *c.* 1660. Image obtained from Lancaster University Library map collection and reproduced with permission.

Plate 2. The jetty at Wardley's Pool, Hambleton. Wardley's and Skippool, not far away on the south side of the Wyre estuary, were the principal ports on the Wyre for many centuries, possibly since Roman times (Crosby, 1998). It is probable that John Ray landed here following his visit to the Isle of Man *c.* 1660. Today Wardley's is a haven for yachts. Photo:© B.D. Greenwood.

Plate 3. Mossy Saxifrage (*Saxifraga hypnoides*) in Ease Gill, Leck not far from where John Wilson found it (Wilson, 1744) Photo: © B.D. Greenwood.

Plate 4. The top portion of a frond of Male Fern (*Dryopteris filix-mas*) collected by James Jenkinson from Yealand in 1774. This is probably the first voucher for a North Lancashire plant record. LSF Gibson MS vol.1 334 James Jenkinson.
© Religious Society of Friends in Britain.

Plate 5. Memorial to James Jenkinson. The inscription reads 'Sacred to the Memory of James Jenkinson of Yealand Conyers who departed this life 15th October 1808 Having nearly completed the 70th year of his age and according to his request was interred here'. Photo: © Norma Heaton.

Plate 6 Ashton Memorial, Lancaster, built on the former Lancaster Moor, *c.* 1910. The Crosfields, Fieldings and Samuel Simpson recorded many interesting plants from here including Parsley Fern (*Cryptogramma crispa*) and filmy ferns (*Hymenophyllum* spp.). Photo: Lancashire County Council, www.lanternimages.lancashire.gov.uk.

Plate 7. Lunecliffe, Stodday. Today Lunecliffe comprises a number of private dwellings and business premises. However Stodday Lodge where the Fieldings lived in the 1830s is incorporated into the present building complex. Photo: © B.D. Greenwood.

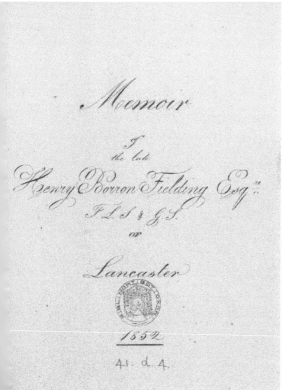

Plate 8. The title page to 'Memoir of late Henry Borron Fielding Esq FLS & GS of Lancaster 1852' MS Sherard 397. Photo: Sherardian Library of Plant Taxonomy, *One of the Bodleian Libraries of the University of Oxford*

Plate 9. No. 1 Winckley Square Preston, the childhood home of Henry Borron Fielding. Photo: © E.F. Greenwood.

Plate 10. Castle Hill, Lancaster where Samuel Simpson practised as a solicitor in the 1830s. Photo: © B.D. Greenwood.

Plate 11. Painting of Common Butterwort (*Pinguicula vulgaris*) by Mary Maria Fielding from vol.1 of 'English Flora 1830–33, MS Eng D 3354. Sherardian Library of Plant Taxonomy, *one of the Bodleian Libraries of the University of Oxford*. The caption to the painting reads '...grew in the field near the deep cut in Ashton, where such a profusion of bog plants grows'. The Fieldings and Samuel Simpson referred to the site as the 'Fairy Field'. It was probably located at approximately SD471592 and not far from the Fieldings' home at Stodday Lodge. Other plants found in the field included Grass-of-Parnassus (*Parnassia palustris*), Marsh Helleborine (*Epipactis palustris*), Pyramidal Orchid (*Anacamptis pyramidalis*), Bird's-eye Primrose (*Primula farinosa*) and Common Cottongrass (*Eriophorum angustifolium*) and no doubt many others of interest. The area today is devoted to intensive grass production.

Plate 12. Rev. Samuel Simpson. Photo Lancashire County Council. www.lanternimages.lancashire.gov.uk

Plate 13. Upper Church Street, Lancaster. Mr and Mrs H.B. Fielding lived at No. 18 Church Street. After Henry Borron's death in 1851 his widow moved to 102 at the top end of Church Street close to St Mary's Priory Church where she died in 1895. Photo: © B.D. Greenwood.

Plate 14. The Lamb and Packet public house, Friargate, Preston where 70 members of a botanical society met, probably in the 1820s. Photo: © E.F. Greenwood.

Plate 15. Stonyhurst College, Hurst Green where Frs John Gerard and Charles Newdigate, presumed authors of the *Flora of Stonyhurst*, taught. Photo: © E.F. Greenwood.

Plate 16. Bruna Hill, Barnacre-with-Bonds, near Garstang from the east. Albert Wilson's family home, Calder Mount, is behind the trees on the right. Garstang and Catterall station was a few minutes walk down the road to the east giving rail access to the rest of the county and nearby to the south on the Lancaster Canal was Henry Fielding's calico printing works that enabled his son H.B. Fielding to live the life of a gentleman. Photo: © E.F. Greenwood.

Plate 17. Knowlmere Manor on the south side of the River Hodder, Newton-in-Bowland. This was the home of Mary Nina Peel. The Heaning, the summer home of J.F. Pickard, is on the north side of the valley almost opposite Knowlmere. Photo: © B.D. Greenwood.

Plate 18. The corner of Beach Road and Clifton Drive North, St Anne's looking south towards the town centre. It is possible that Atherstone House, where Charles Bailey lived, was one of the villas shown here built by William John Porritt at the end of the nineteenth century. They are now within a designated conservation area. Behind the houses are the Ashton Gardens, formerly St George's Gardens. These had been conceived as early as 1875, two years after St Anne's started to be built (Shakeshaft, 2008). When Charles Bailey lived in St Anne's (c. 1901–1010) they were little more than levelled dunes. Similarly the dunes on the north side of Beach Drive were also levelled and the new street pattern laid out but used for chicken pens before being developed. It was this area that provided a fruitful hunting ground for Charles Bailey and is illustrated in his paper on evening primroses (Bailey, 1907b). There is a plan of the town dated 1899 in Haley (1995). Photo: © B.D. Greenwood.

Plate 19. The Parish Church, Churchtown, near Garstang. This is the last resting place for Henry Borron Fielding and his wife, Mary Maria. In the adjacent grave lies Henry's sister Henrietta and her husband, the Rev. John Clay who may have written the 'Memoir' of H.B. Fielding. Inside the church is the grave of H.B. Fielding's father, Henry and other members of the family. Also buried in the churchyard is R. Sharpe France. Photo: ©E.F. Greenwood.

Plate 20. The Floras of North Lancashire. Photo: © B.D. Greenwood. Cover images are reproduced by kind permission of the Regional Heritage Centre, Lancaster University for M.J. Wigginton, *Mosses and Liverworts of North Lancashire* (Lancaster, 1995); J. Matthew Leedal for L.A. & P.D. Livermore, *The Flowering Plants and Ferns of North Lancashire* (1987) and P.P. Abbott for *Plant Atlas of Mid-west Yorkshire* (2005).

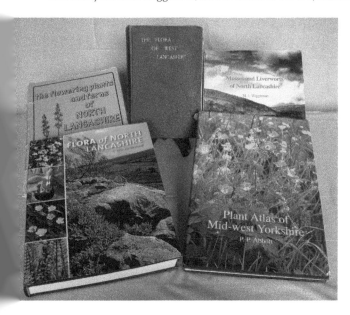

Plate 21. Charles Bromley-Webb. Photo: ©Margaret Bromley-Webb.

Plate 22. George Wade Garlick in March 1982. Photo © C.M. Lovatt.

Plate 23. Dr John Hodgson with his mother, Eileen and daughter Claire by the R. Derwent, Hathersage, Derbyshire, 1989. Photo: ©Janet Hodgson.

Plate 24. Len Livermore left, Jeremy Steeden centre and Pat Livermore right hunting for ferns in Perthshire July 1977. Photo taken by C.F. Steeden. © N.J. Steeden.

Plate 25. Dr Jennifer Newton with her husband David in Aughton Wood in the Lune Valley, 2007. Photo: ©Dr Jen Cars.

Plate 26. Bernard Oddie taken in 1997 on the island of Islay. Photo: © Dr Richard Gulliver.

Plate 27. Liverpool Botanical Society on a field trip to Murdishaw in 2005. Vera Gordon is third from right. Photo: © Wendy Atkinson.

Plate 28. C.F. (Derek) Steeden in western Crete, May 1982. Photo: ©N.J. Steeden.

Plate 29. St Michael's Church, St Michael's on Wyre. Phipps J. Hornby and John Moss are buried here. Photo: © B.D. Greenwood.